In-Flight Entertainment & Connectivity

In-Flight Entertainment & Connectivity

The history and current state of IFEC in Commercial Aircraft and Executive Jets.

By

Rory Briski

ISBN: 978-0-9827921-3-1

Published by: www.SpiralJourney.com

DEDICATION

This book is dedicated to my wife Astara, without whom, none of this would have been possible.

USE OF 3rd PARTY PICTURES AND DRAWINGS

Throughout this book I use a vast quantity of pictures, drawings and illustrations. Some of which I have created myself, some of which reside in the public domain or have Creative Commons Licenses, some are part of clipart collections with permission to use in works such as this, and some of which have copyrights belonging to 3rd parties. Insofar as possible, I have obtained permission to use this 3rd party material.

However, if in the course of editing, if one was missed, and you are the copyright holder, and would like it removed or credited differently, please contact me and I'll revise/edit immediately. No infringement of any kind is intended.

Thank you.

ACRONYMS

There are a lot of acronyms used when dealing with aircraft systems. So, at the very end of this book you will find a handy acronym list.

DISCLAIMER

Within this book I use a variety of numbers for such things as quantities, fees, subscription prices, costs, etc. So, please note that the numbers and fees listed herein were reasonably accurate based on general industry practices, and publicly available information, at the time they were being written about. All data listed is perishable and subject to change without notice.

Table of Contents

Preamble

In-Flight Entertainment has been around since the beginning of flight. The "forms" of entertainment may have changed, but the underlying principals have remained the same.

In this book, I endeavor to convey to you the essence of the history and present day systems and technologies that have shaped the In-Flight Entertainment & Connectivity (IFEC) landscape.

The book is divided into four (4) main sections: This brief beginning section, a section on Air Transport Class Aircraft (typical commercial airline aircraft), a section on Business, and Corporate aircraft, and lastly a section on VIP & VVIP aircraft.

Each section builds upon the last, so it is preferable that you read from beginning to end.

With regards to the "Connectivity" part of this book, my focus here is on the functions that the connectivity enables, and not so much on the underlying technology. Such as which antenna type is better, single feed-horn, multiple feed-horns, phased array. Or what frequency is best to transmit on: 1.5GHz, 16GHz, 33GHz, etc.

The focus here is on the application of the enabling connectivity technology as opposed to the technology itself. Perhaps I'll write another book with focus on that piece, but who knows when or if that muse will strike...

What is In-Flight Entertainment (IFE)?

What are your thoughts?

What does IFE mean to you?

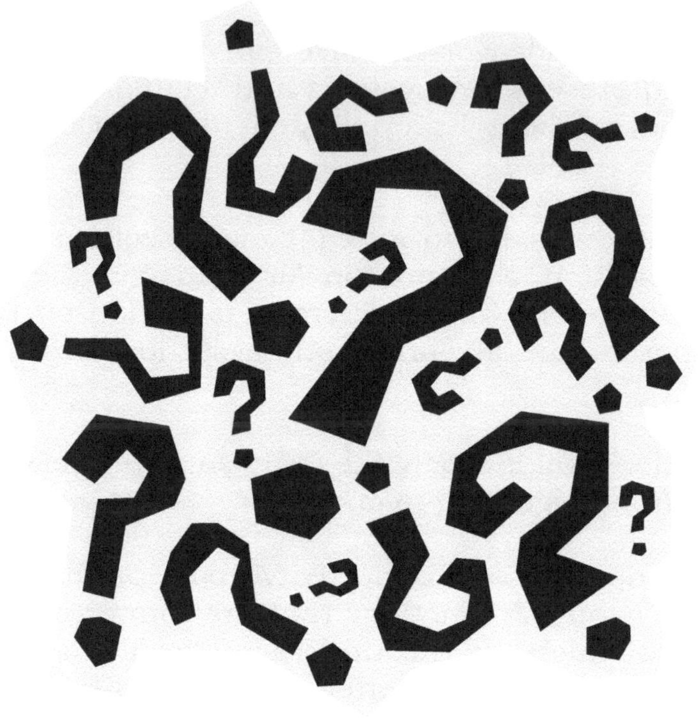

I want you to really think about that for a minute before turning the page. Before you travel, do you think about it? When you make your reservation do you consider the IFE options? Does having on-board music, movies, laptop power or an Internet connection impact your traveling decisions?

Okay, now that you've thought about it, I will give you a few of my thoughts on the subject.

IFE is a very BROAD category. It is more than just electronic systems on-board the aircraft.

Content is just as important if not more important for the end user. It's all well and good to have a wonderful 15" High-Definition (HD) Video Monitor in the seat, but if there are no HD movies to show on it, or really bad movies, it has little to no ability to keep the passengers entertained. In this regard, content is king.

But then again, if you have great content, but nothing to show it on, you have nothing for the passenger either. So, how about for now we call them both equal. Okay? Good.

But wait a minute, what if I don't want to watch a move. I want to get some sleep. So what about sleeping? Is sleeping IFE?

On September 20, 2011, Finnair had a flight from Helsinki to Singapore that was a dedicated "Angry Birds" flight. Complete with people in Angry Bird costumes and with Flight Attendants handing out portable devices pre-loaded with Angry Birds games. I understand that was every bit an In-Flight Entertainment experience.

IFE Concepts

IFE is not a single entity. IFE is an EXPERIENCE.

You need to look at IFE as a whole, and not just bits of electronics installed around the aircraft cabin.

Generally speaking, IFE is ANYTHING that distracts the passengers and allows them to pass the time in the air more quickly. Typically this is music or other audio programming, video programs and movies, and of course games. And for a growing number of people, reading, working and surfing the internet are key elements of their In-Flight Entertainment experience.

IFE has several distinct parts, and I'll be covering each in differing amounts of detail as you progress through the book.

- Hardware: These are the electronic bits and cabling scattered throughout the aircraft and installed into the passenger seats.
- Software: This is the operating system and applications that make all the hardware play nice.
- Programming: These are the Graphical User Interfaces (GUI) elements and system configuration tools.
- Content: As I mentioned previously, these are the movies, books, music, etc., that distract the

passenger and provides them with viewing, listening, and entertainment pleasure.

IFE also has a couple of broad categories in that it can be provided by the airline or brought on-board by the passenger.

Typically the airline, corporate or private flight departments provide items that include:

- Video Screens (Overhead/In-Seat/Portable)
- Audio/Video Distribution Systems
- Internet Access/E-Mail/Telephony
- Satellite Radio/Television
- Laptop/Tablet Power
- Video Games
- Decks of playing cards, coloring books
- Pillows & Blankets
- Other Similar Items

Typical items that passengers bring on-board include:

- Books/Magazines
- Board games
- Electronic (Gameboy type)
- Manual (Cards, Dice, etc.)
- Laptop Computers, Netbooks, Tablets

- MP3 Players, iPods, Smart Phones
- Personal Video Players (DVD, Blu-Ray or Hard-Drive)
- Other Similar Items

So, as you can see there is a fair amount of diversity and a fair amount of commonality between what passengers bring on-board and what is typically provided by the airlines and flight departments.

AIR TRANSPORT AIRCRAFT

Okay, we are going to start this journey into IFE land with the Air Transport class of aircraft. These are typical airlines like: United Airlines, Lufthansa Airlines, Singapore, Virgin, etc. Remember, we'll discuss Business, Corporate, VIP and VVIP aircraft in later sections.

So, what do these airlines, and other airlines like them, all have in common? (What's that? You thought I'd only be giving answers and not asking questions? Well, you can count on me to pose questions like these fairly regularly. It should keep you on your toes and paying attention.)

What they all have in common is that they all use the same basic aircraft manufactured by Boeing and Airbus.

Yes, there are some regional jets by Embraer, Bombardier and others, and many turbo-props with quite a few vendors, most notably ATR. But for the most part, what I'll be discussing here are aircraft that traditionally have IFE systems installed.

All aircraft are basically the same; they are a long hollow tube, with wings and engines. That is what the aircraft manufactures sell, big hollow tubes.

They may have different flight characteristics, payload capabilities, speed, fuel burn, etc., but they are all essentially hollow tubes that need to be fitted with nice interiors. Oh, and engines... The engines are usually purchased separately from the aircraft as well. But, for this book I'm focusing on the interior and the IFE system.

So, the airline bought the aircraft and the engines and now it is up to each individual airline to outfit these tubes with interiors that they believe will be attractive to their travelling target market.

The airlines pick the seats, baggage/stowage bins, color schemes, carpeting, set the seat-pitch and most importantly, decide which IFE options that they want to offer their passengers. Now, some of these choices can be severely restricted by the aircraft OEM for a new aircraft delivery, but in a retrofit situation, the airlines can pretty much get whatever they want.

Air Transport IFEC Market

Just a short bit here on the overall In-Flight Entertainment & Connectivity (IFEC) Market. The IFEC marketplace is dynamic and continually growing. Even during the recent economic downturns, airlines continued to upgrade systems in order to lure in passengers.

Some IFE historical expenditure numbers:

- 2005 = $1.8B USD
- 2012 = $2.9B USD

That is certainly only a modest growth from 2005 to 2012 but still a significant amount of money being spent on IFE.

Looking ahead at the forecasted IFEC expenditures for 2020 we can see significant growth at around $7.5B USD. This has hardware and content growing at about 13% and connectivity growing at over 28%. With connectivity revenues alone reaching around $2.4B USD per year.

This looks like a pretty lucrative marketplace over the next several decades.

Aircraft Types

Commercial aircraft basically come in two flavors, Wide-body (Twin-Aisle) and Narrow-body (Single-Aisle).

Wide-Body (Twin-Aisle) Aircraft

These aircraft typically have a very long range (8 to 14+ hour flights) and they are typically "Over Water" routes.

Alright, a quick break here for a moment. You may have noticed that I use the words "typically" and "basically" fairly often. That is because, like with most things, there are rarely absolutes. For example, a B747 is a wide-body aircraft, typically long-haul over water, but All Nippon Airways (ANA) uses these big aircraft for commuter flights between Tokyo and Osaka in Japan. So, if I miss a "mostly" or "typically" or "often" in a statement somewhere, please keep in mind that there are probably exceptions. Thank you and I now return you to your regularly scheduled reading.

Okay, wide-body aircraft typically have the most IFE options of any Air Transport class aircraft type. These options run the full gamut of:

- Broadcast Audio
- Audio-on-Demand
- Broadcast Video
- Video-on-Demand
- Satellite TV
- Internet TV
- Internet Radio
- Satellite Radio
- Video Games
- E-Mail
- Web Browsing
- Language Courses
- Self-Help Courses
- In-Seat Exercises
- Traditional Books and Magazines
- Electronic Books and Magazines
- Satellite Phone Service (in-seat)
- Personal Cell Phone Services (GSM Phones)
- In-Seat Power (for laptops, tablets, phones)
- Traditional On-Board Shopping
- Electronic Shopping

- Traditional Duty-Free Shopping
- Electronic Duty-Free Shopping
- Interface to passengers' iPod
- Ability to watch a movie on a large screen in the seat while reviewing meta-data, text and internet content about the movie on a smaller in-seat screen.
- News, weather, sport scores, connecting gate information
- Moving Map
- Moving Map with position of interest (POI) data
- Reading light controls

There are probably a few other things that I am forgetting but as you can see there are a lot of IFE options typically available in a wide-body over water aircraft.

Narrow-Body (Single-Aisle) Aircraft

These aircraft typically have a short range (2 to 5 hour flights) and they are typically "over land" routes. (Also, when talking about narrow-body aircraft, Regional Jets fit into this category. Although their IFE offering has historically been slim to none. But that's changing a bit and I'll get to that later.)

With regard to IFE options, historically, narrow-body aircraft had little to no IFE. Sometimes they had audio, or maybe audio and overhead video but that was about it. However, that was then, and this is now. Currently you can find narrow-body aircraft with no IFE at all, and some with the same full range of capabilities as seen in their larger cousins the wide-body aircraft.

Typically though, narrow-body aircraft have a minimal compliment of IFE options. Because, the longer the flight, the more "distracted" you need to make the passengers. And narrow-body aircraft just don't spend that much time in the air on a given flight.

Long Haul versus Short Haul Aircraft

This distinction is typically split along the same lines as Wide & Narrow-body aircraft. Wide-Body are normally Long-Haul and Narrow-Body are normally Short-Haul.

However, unique routes require unique solutions, as in my example about ANA earlier.

Additionally, some airlines dedicate aircraft to specific routes, some don't. That is, they may take six aircraft from their entire fleet, of say 100 aircraft, and have them only fly between two or three airports. That way, they can match the IFE and other amenities and aircraft characteristics to the specific route the aircraft flies.

If they do not dedicate aircraft to a specific route, then all aircraft in the fleet that could be routed to fly a given route must be properly equipped.

A number of years ago airlines ran the New York to Washington D.C. shuttle with aircraft equipped with in-seat phones. Compared to other routes around the USA, it had extraordinarily high in-seat telephone utilization for voice and email traffic.

Also, given today's marketplace, airlines are offering IFE on smaller and smaller aircraft in order to gain market share and thereby increase passenger revenues.

Airline Branding

Jet Blue mastered this approach when they launched their airline.

Without launching with IFE, and a very advanced IFE at the time, their launch would have been just another hoo hum low cost carrier entering the market.

But branding their launch with Satellite TV and In-Seat Video brought them incredible market buzz and TV coverage and was a brilliant move.

Long Haul Flyer Segmentation

Air New Zealand, one of my favorite airlines, did a study a while ago that looked at typical passengers on long haul flights and split them up into five interesting categories. Which one, or more, of these categories do you belong to?

- Positivists (10%)
  - Fidgety and excited
  - Highly involved in the flight and with the romance of travel
- Socialites (29%)
  - Social
  - Highly involved in the flight and with those around them
- Cocooners (17%)
  - Entertain themselves
  - Wish to zone out and not be highly involved in the flight
  - Chuckle at Positivists and families
- Territorialists (26%)
  - Predominantly frequent flyers
  - They know exactly what their space entitlements are
  - Have a high expectation of service

- Disengaged (18%)
  - The flight is just a means of getting somewhere
  - Want to keep to themselves

So the question becomes, how do you structure an IFE offering that will appeal to all five of these groups? Or at least appeal to them in some way so as to make their flight experience more of a pleasure than a pain.

This is where diversity, freedom of choice, and airline competition takes place, by trying to satisfy this diverse crowd of flyers.

Perceptions of IFE Importance

Typical Passenger Perspective

From the long haul segmentation we saw that 17% of the travelers preferred to entertain themselves, while the other 83% all wanted, by varying degrees, to be entertained.

So generally speaking, I would say that from a passenger perspective, IFE is VITAL to their travel experience. Being late is one thing, but if their IFE doesn't work they are VERY vocal about it, passengers expect to be entertained in-flight. And finally, "Misery Loves Company".

Misery loves company? What is that doing here? Well, in the early days of IFE there were big projection screens in front of the cabin and everyone watched the same movie. There were no in-seat video monitors, just big screens like you would find in a movie theatre on the ground.

If the projector failed, which they frequently did, everyone in that cabin was out of luck, no movie for anyone. People were generally unhappy but not a big deal. Now, along comes in-seat video.

With in-seat video imagine everyone is watching a movie of their choice, having fun and suddenly your monitor goes blank, and only your monitor goes blank. Everyone around you is still enjoying themselves and you have a 7 hour flight left to go with no IFE. You are going to be a lot madder than if everyone else in the cabin lost their entertainment to, hence the saying, "Misery Loves Company".

Some things are just more easily tolerated on an individual by individual basis if everyone is inconvenienced.

Also keep in mind that air travel is frequently, cramped & uncomfortable, which means shorter tempers, which could result in air rage. (Like road rage, but in the air.)

Passengers have come to expect, in flight, the same amenities that they have in their homes and cars.

Typical Airline Perspective

For an airline, all aircraft are hollow tubes with wings and engines. The cabin interiors and IFE are where they can differentiate themselves and attract passengers. This is because all airlines have access to the same basic aircraft from the aircraft manufacturers. No aircraft Original Equipment Manufacturer (OEM) sells an aircraft platform exclusively to only one airline. Where they compete is with what's inside the aircraft. And to a lesser extent with personnel and services of course, but IFE plays a key role in that differentiation.

So, for an airline, IFE is a necessary evil. If an airline could operate without IFE I'm sure they would, some small airlines actually do. IFE is an enormous expense, in both purchasing costs and operating costs. The costs of which include fuel costs to carry all the weight of the IFE system around.

What airlines keep trying to figure out is: IFE is an expense, how do we minimize it? Or better yet, how can we make money from it?

More than ever before airlines are trying to find ways to make money from their IFE systems.

Here is a short list of the more common types of revenue streams airlines are typically looking at:

- Duty-Free Sales

- Advertising
- E-Mail
- Web Surfing
- Phone Usage

Not only charging passengers for these services, but charging the advertisers as well. Advertisers pay the airlines millions of dollars to have their products shown to this typically affluent and captivate audience. If someone doesn't like the commercial, they can't exactly get up and walk out.

Typical OEM Perspective

The typical OEM, here that would be the aircraft manufacturer, also views IFE as a necessary evil. It is a huge pain for the OEMs to deal with all of this custom IFE when they are trying to build a bunch of aircraft all exactly the same. And every time you add IFE into the mix it has an impact on their production line.

And things that impact the production line by causing slowdowns, rework or work stoppages, cost the OEMs thousands of man-hours in lost productivity.

The OEMs are always looking for ways to minimize the impact to the production cycle. They are also trying to more efficiently manage changes to the program or functional requirements. Do they make IFE a basic part of the aircraft? Do they allow the airlines to pick whichever IFE company and system they want? Some middle ground? Charging for changes?

The OEMs have certainly moved away from the past practice of allowing the airline to choose whatever IFE vendor they wanted. These days an airline is lucky to have any choice at all. For most OEMs, and most wide-body aircraft types, it is down to a choice between only two companies for a line-fit IFE system, Thales and Panasonic.

It is possible, on a few aircraft types, to pick a third company, but the OEM would typically charge the airline millions of dollars to do so. Not very promising for young, up and coming IFE companies. But, for them, the retrofit market is the jewel, or brass ring, or gold ring that is often their target.

But I digress, back to OEM line-fit of IFE systems. Standard IFE systems and the offering of standard option packages to go with them seems to be all the rage now. The OEMs pre-qualify as much equipment as they can, or is practical, and offer only those pre-qualified systems and system options to the airlines.

Historically, Boeing used to offer more choices than Airbus, but now it seems that they both offer only the same two vendors.

The trend here is for pre-wired aircraft and requiring the IFE vendors to use the pre-defined OEM wiring scheme. The A380 and B787 being the latest examples.

But even with a common wiring plant, if a 3rd party came in and said their system could fit into that wiring scheme, I understand that neither Boeing nor Airbus would allow them on-board without paying millions of dollars in "development research" costs to ensure that they indeed did fit properly.

History of IFEC

Our foray into the skies began around the early 1900's. (I'm not going to get into the argument of who was first to get a plane in the air. Regardless of who you think was first, it happened around the same time.)

Before we get into the meat of this, a quick question:

When do you think the FIRST In-Flight movie was shown?

Go ahead... Think about it a bit... I'll wait...

Most people I ask say in the 1950's. A few have said in the 1970's and some say in the 1940's... They are all shocked when I reveal the answer.

The answer is, 1925. Yes, just a short 25 years or so after we took to the skies for the first time, the first in-flight movie was shown. It was in a WWI (World War One) Handley-Page bomber that was converted to carry commercial passengers. They played the black and white film "The Lost World" during a 30 minute flight near London, England.

Having said that, there is a small cadre of folks that claim the first was in the USA, in 1921. They claim it was on an Aeromarine Airways flight around Chicago, IL and showing the film "Howdy Chicago". So either 1921 or 1925, still, a very long time ago...

Anyway, here are a few more "firsts" for you to think about. While mostly related to Air Transport, there are some Executive Jet firsts listed as well:

- ✈ 1932: First Inflight Television
    - Planes flew a lot lower and a lot slower back then. The antenna was in the nose cone.

- 1961: First Feature Film Inflight
- 1961: 16mm Films on 30inch Reels
    - These were huge film canisters that were threaded throughout all the projectors in the aircraft, starting up front and snaking its way to the back. So if you were up front and you missed something, you could run to the back before the "film" threaded through the projector back there and possibly be in time to see what you missed.
- 1962: Pneumatic Headsets Introduced

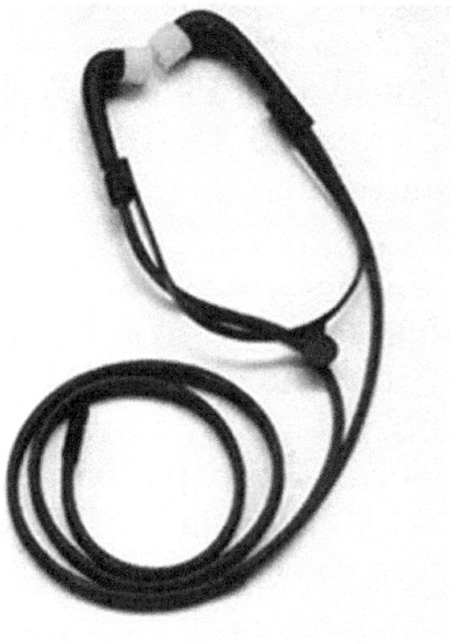

Figure 1 - Pneumatic Headset

That's right, for 37 years, from 1925 until 1962, when they played a movie or TV show on-board an aircraft, the audio was played through the cabin Passenger Address (PA) system. When it was time to watch the movie EVERYONE had to listen to it, whether they wanted to or not.

On a side note, some of the early multiplexed IFE systems, with in-seat headsets, used a test tone sent through the system for its built-in-test (BIT). That was a troubleshooting aid to have the system help to self-diagnose any problems.

Well, it just so happened that a 1968 movie called "Ice Station Zebra" had a scene in it that had enough pink noise generated in their sound track to make the system "think" it should go into self-test mode.

In a very suspenseful scene, the cabin is dark except for the flickering of the movie projector, all the passengers were on the edges of their seats, critical moment in the move and... The movie screens go blank, the cabin gets pitch dark, the flight attendant chimes start ringing, all the cabin reading lights start flashing, and general panic ensues... It took a few moments but the stewardesses, as they were called then, managed to get everyone calmed down.

The IFE BIT (Built-In-Test) and BITE (Built-In-Test Equipment) has come a long way since those days so things like that "shouldn't" happen again. At least not from the same cause anyway.

- 1971: 8mm film cassettes
  - Huge upgrade to reliability and portability for the airlines when they switched to these systems. In fact, there are still aircraft flying today that still use these old tapes.
- 1975: Atari Hand-Held Video Games used on-board
- 1978: VHS Video Tape Systems used
  - Larger tape format than 8mm. Studios released a lot of content on these tapes.
- 1979: Electronic Headsets in First Class
  - They were fairly large by today's standards but they sounded a 1,000% better than the pneumatic headsets.
- 1984: Inflight Telephone Service
  - Phones that were built-into the seats. These had a connection charge and a per minute fee. It varied quite a bit while the airlines and service providers tried to find the sweet spot, but the average domestic call seemed to hover

around a \$2.50 setup fee and \$2.50 per minute usage charge. And International (SATCOM) calls at \$10.00 setup and \$10.00/minute.

- 1988: In-Seat Video System (2.7" CRT)
  - Yes, that says CRT, as in Cathode-Ray-Tube. As in 15,000 volts a few milli-meters (fractions of an inch) from the back of your head. (The monitor in the back of "your" seat was used by the passenger behind you, and you watched the one in the seat in front of you.)
  - It was certainly novel at the time, your own personal video screen and a selection of broadcast video channels to choose from.
- 1989: Noise-Canceling Headsets were introduced
- 1989: Fleet wide In-Seat Video in First Class (LCD)
  - The introduction of Liquid Crystal Display (LCD) In-Seat Monitors effectively killed off the fledgling CRT market, what with its lighter weight, larger screens and presumably safer installation.
  - Japan Airlines (JAL) was booked a full four (4) months in advance on flights from the US to Tokyo when they introduced this service. That was an unheard of increase in the passenger load factor, and market share.
- 1991: Seat-Back video in all classes
  - It didn't take long for airlines to start offering In-Seat Video screens to Coach/Economy class passengers to try and increase their market share with that segment of passengers.
  - Needless to say these were long-haul, wide-body, over water aircraft and routes.
- 1993: Global FAX via Satellite
  - Remember FAX machines?
- 1996: In-Seat Power (15Vdc)
  - Laptop power ports started to be installed. No more having to worry about dead batteries. All

you needed was a typical laptop car charger/adapter.

- 1996: Live TV (via Satellite) was deployed
- 1996: The First Dolby 5.1 Surround Sound System Installed
  - This was on an MD-87 VIP aircraft delivered from the Jet Center/Van Nuys.
- 1996: Pioneer and Gulfstream validate the use of DVD players on Executive Jets
- 1997: Interactive AVOD, Full Cabin
  - AVOD = Audio & Video On-Demand
  - Passengers could watch what they want, when they want it. Airlines started moving away from the "broadcast video" solutions and to this new paradigm.
- 2000: Live Satellite TV, Full Cabin, in-seat video
  - Now everybody gets to share in the satellite TV goodness.
- 2000: Fleet wide, All Cabins, Portable DVD Players
  - Handing out (for rent or free) portable DVD players gave the passengers even more freedom in what they wanted to watch.
- 2000 In-Seat Power Installed (110Vac 60Hz)
  - No more need for a "car charger" type adapter, just plug in your normal AC power cord and you were good to go.
- 2001: First E-Mail Transmission via a dedicated IFE system
  - This was via a keypad built into the IFE handset at the seat. On one side of the handset was the phone and the other side had the QWERTY keyboard and the IFE controls.
- 2002: First SMS Text messaging via seat-back display
  - Seat-back touchscreen with a virtual keypad.
- 2003: First portable AVOD units deployed
  - AVOD content on hard-drives instead of on DVDs

- 2004: Launch of Connexion by Boeing (CBB) with Live Internet
    - Finally, real live high-speed internet access that didn't cost an arm and a leg and was actually useful. My favorite IFEC thing ever... Okay, this is really tied with In-Seat Power as my favorite...
- 2004: Proof of concept PICO cell inflight (CDMA) for In-Flight use of Cell Phones
    - Use your own, personal, cell phone to make and receive phone calls while in-flight.
    - I'm writing this in 2014 and only a handful of airlines and countries permit you to use your cell phone in-flight and only if they have this technology installed. The USA currently does not permit their use.
- 2005: Four channels of live TV over CBB to passenger notebook computers
    - What "can't" you do once you have a high-speed internet connection in the air?
- 2007: High-Definition (HD) Video systems and monitors deployed
- Hummmm... Since 2007 what's new?
    - Sat TV? Nope. It's gotten better but it's not new.
    - Internet Access? Still no. It's faster now but not new.
    - Cell Phones? No again.
    - Portable IFE? Nope. We have Tablets now instead of DVD players and other purpose built units but the concept is the same.
    - Portable Electronic Devices? Nope.
    - Wireless-IFE? Not really, I'd say that started when internet access was first permitted.

There are definitely NEW things coming, but you'll have to wait a bit to find out what...

Commercial Air Transport: Traditional System Types and Functionality

The focus here is on the "Entertainment" side of things, and not so much the Cabin Management System (CMS). The CMS is what controls the lights and Flight Attendant Call and other bits around the cabin.

As I talk about system types I primarily break them out into two categories based on how they are typically wired.

- Star-Wired
- Daisy-Chained

Star-Wired

Star-wired systems typically have:

- Complex Engineering Drawings
- Lots of Wire
- Labor Intensive Installation
- Fault Tolerant Downstream
- Single-Point Failure Upstream

The next two figures (Figure 2 & Figure 3) show the star-wired concept.

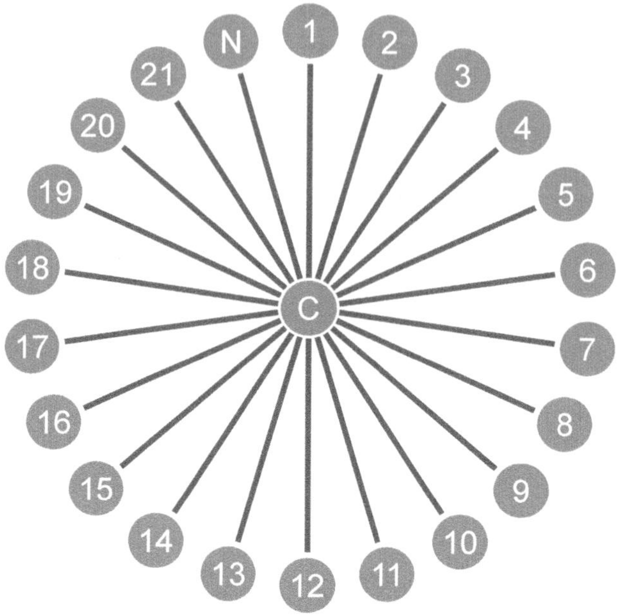

Figure 2 - Star-Wiring Concept

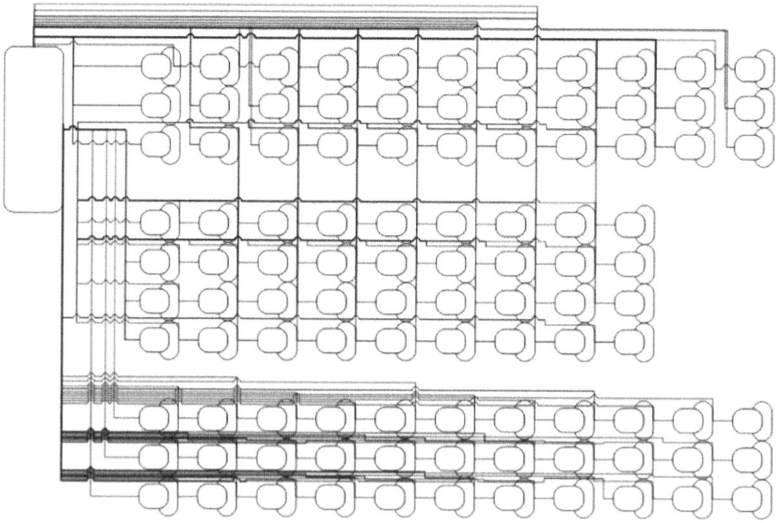

Figure 3 - Star-Wiring

The "C" in the middle of Figure 2 represents the head-end. This is where all the movies are and where the "brains" of the system are located. The numbers circling it represent the seats.

Figure 3 is a simple pictorial that shows the wiring clusters from the seats to the head-end in one small part of an aircraft cabin. Now imagine 400+ seats all wired up to the front of the aircraft like this. Yikes! What a nightmare.

To keep the cables as small as possible these were made with teeny tiny wire gauges, something on the order of 32awg or even smaller. These were very prone to breakage, crosstalk (interference between channels), and generally a pain to work with.

Then, after installing a few of these systems, they started to get smarter and instead of wiring every seat individually to the head-end, they wired each seat "group" to the head-end, and had a little junction box in the seat where the signals would be tapped off and routed to each passenger location. That eased the installation, wiring and system weight burdens to an almost acceptable level.

Now, prior to having multiplexed type signals, which I get into more depth on shortly, these systems were all analog. Imagine for every channel of audio that you want to listen to, you had four wires (2 for left ear and 2 for right ear). And these wires were routed to every passenger in the aircraft. Now, if you want 10 channels of audio, you route 40 wires (plus their shields)... Everywhere...

As you can imagine, a very heavy and inefficient proposition. Also, as more people listened to the same program, the volume would drop because of the added load on the line.

Daisy-Chained Wiring

Daisy-chain wiring is where you have electronic boxes that are wired together in a string, one after the other.

Historically there are several different types of daisy-chain systems including:

- Multiplexed
- Analog (RF)
- Hybrid Analog/Digital
- Digital
    - Ethernet
    - Proprietary

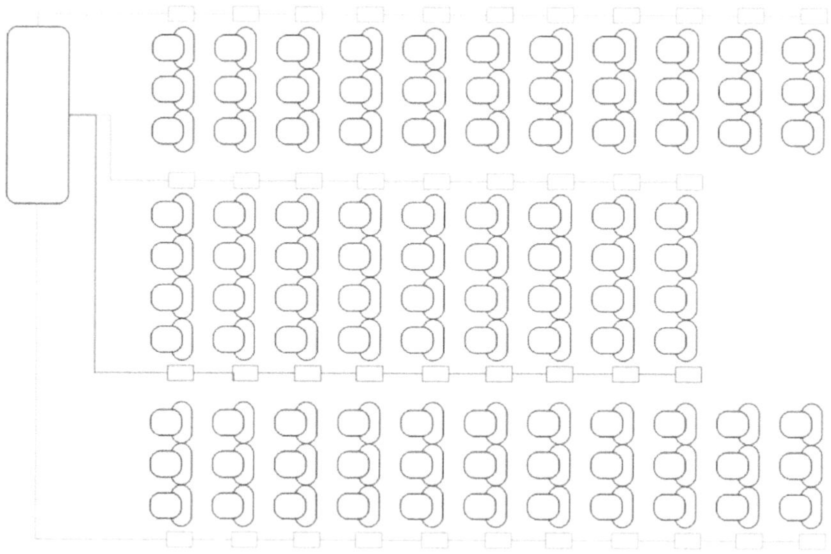

Figure 4 - Daisy-Chain Wiring

Figure 4 - Daisy-Chain Wiring shows a small section of a wide-body aircraft with a daisy-chain cable plant. The quad seat groups in the center are either fed by one seat box per seat group, handling all four seats, or by two seat boxes, each handling two seats. It just depends on the vendor and which system they are offering.

Daisy-chain systems typically have:

- Simple to wire installations
- Simple engineering drawings
- Most IFE systems have been installed in this or a similar manner.
- Sometimes the systems are wired more in a T-shape where the cabling comes into the middle of the column and then daisy-chains forward and aft. This is still considered a "daisy-chain" system.

Multiplexed Systems

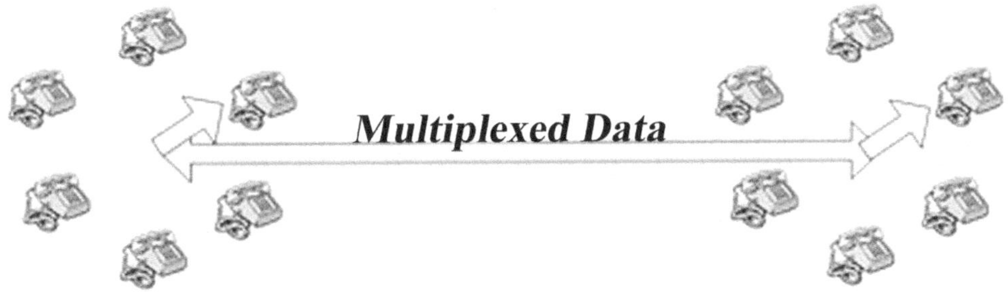

Figure 5 – Time-Division Multiplexing

Prior to the wide spread use of Ethernet, multiplexed systems were number one.

These systems typically used methods like adaptive delta modulation techniques that allowed for:

- The information to be sampled and time sliced
- Using Time-Division Multiplex
- Some systems used Frequency Division Multiplex instead

MUX - WHAT IT IS!

Before I move on to the other system types, I'd like to dwell on Multiplexed (MUX) systems for a moment. These were a real turning point for the IFE industry so a bit more history and how they function may be in order. *I originally wrote most if this next section for an article that was published in AVION Magazine in 1989. AVION was the official publication of the World Airline Entertainment Association (WAEA). (An organization that is now called the Airline Passenger Experience Association (APEX)).*

It began in the early 1960's, airline passengers were able to finally select and listen to their own individually selectable audio programs and the world hasn't been the same since. The B707 and DC-8 were two of several types of aircraft that hosted this revolutionary innovation.

A primary piece of equipment in use at that time was the CAM-1 Tape Reproducer developed by Transcom. This unit used large tape cartridges to house the audio tapes. The cartridges were unique in that the tape was not spun upon a reel but was folded back upon itself repeatedly. This made the tape look like a very long snake coiled back and forth within the shell of the tape cartridge.

The audio signals were fed into a distribution system manufactured by EECO Inc. This system consisted of an audio amplifier, many junction boxes and of course a PCU (Passenger Control Unit) at each seat. The audio amplifier took the outputs from the CAM-1, amplified them and sent them to the passengers via the junction boxes. Each mono-audio channel required two wires to be routed to the junction boxes. If six channels were made available then twelve wires were required to be fed throughout the aircraft. If ten channels were available the twenty wires had to be fed throughout the aircraft. The junction boxes in turn split the wires again to feed each Passenger Control Unit within its respective seat group.

The PCU contained a volume control knob and a channel select knob. The audio was provided by two small speakers within the PCU. The sound from these speakers was transferred to the passengers' ear by hollow tubes called pneumatic headsets.

This method of audio distribution was a very good application of the technology available at that point in history. The frequency response was good, the reliability was fair and the passengers were happy to have the ability to choose their own audio programs.

However, as the novelty of In-Flight Audio wore off the passengers began demanding more. The design of the Hard-Wired System caused many cross talk problems. That is when you listen to one audio channel but can also faintly hear another audio channel in the background. Also, as more and more passengers listened to the same audio channel, the signal level was reduced to the point that it was inaudible in the rear of the aircraft. The many thousands of tie points within the system also caused degradation in the audio signal. Additionally, if an airline wanted to change its programing from mono to stereo, then major wiring changes had to be done to the aircraft. So while the system operated as expected and had an overall effect of pleasing the passengers, it wasn't without its problems.

MUX or Multiplex systems have been in common use since the mid 1940's. The first widespread commercial use was in telephone systems to increase the number of calls that could be placed on one telephone line. The MUX system is easiest to describe in its most basic form by referring to their early use by Bell Telephone Company. As shown in the accompanying drawing (Figure 6), the multiplex system can be visualized as several telephones connected to only one line through two switches, one at each end of a single pair of wires. As this switch is turned each telephone is sequentially connected to its corresponding telephone on the other end of the line.

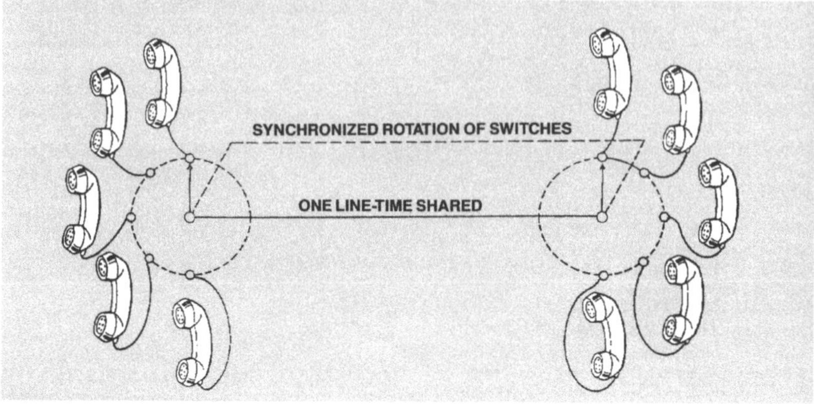

Figure 6 – Simplified Multiplexing Using Telephones

A more formal definition would be that multiplexing is a technique used to time-share a common carrier being fed several different audio or data inputs. Each time the switch engages a given telephone line it is said to be "sampling" the audio on that line. By sampling we mean cutting out a small slice of the audio signal. This "slice" is then labeled for later identification and then sent down the line to the receiver to be demultiplexed and sent to the appropriate telephone.

In 1970 Pan Am became the first user of a multiplexed audio distribution system in the airline industry. This system was installed on a B747 and was manufactured by Telephonics Corp. Unfortunately, it did not fare as well as the Hard-Wired systems in use on the Narrow-Body aircraft. The MTBF (Mean Time Between Failures) of the MUX system was a mere five flight hours compared to 85 flight hours of the EECO system. The units that make up a wide body MUX system and their interrelationships are shown in Figure 7.

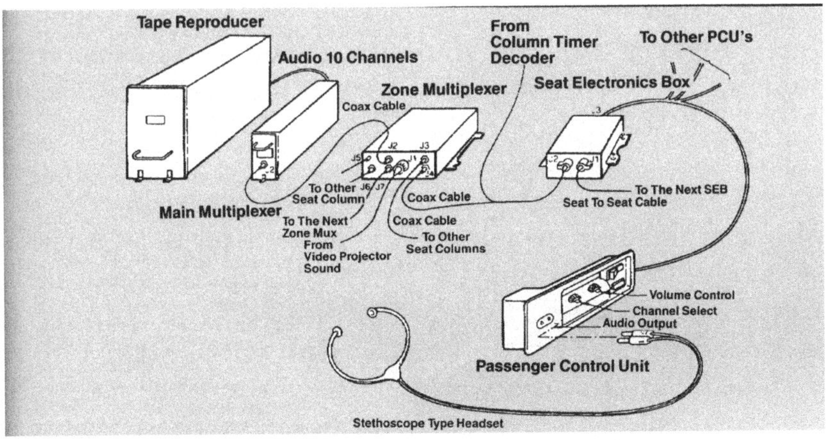

Figure 7 – PES System Simplified Diagram

One of the driving factors for MUX on these Wide-Body aircraft was the ability to multiplex control signals for the reading and call lights. Due to the twin aisle design and the high ceilings of these aircraft it was impossible for passengers to control these functions from their seats if the controls were put into the overhead as on a B707. Having to run wires throughout a B747 to accomplish service and audio would run into about 1,556 pounds. A MUX system accomplishing the same purpose would weigh only 580 pounds. As weight is gold to an airline, MUX had at least one clear advantage over Hard-Wired. These same concepts were used on the DC-10 in 1973 with American Airlines and United Airlines being the first customers of that type and on the L-1011 in 1975 with TWA, Eastern and Delta kicking it off.

As technology progressed, some of the old cordwood circuit designs gave way to Hybrid integrated circuits which were a major factor in raising the Telephonics MUX system MTBF to an unheard of 100 hours.

Telephonics was not the only player with a MUX system, Hughes Aircraft Corp. also offered its version of a MUX system. Hughes had a distinct advantage over Telephonics in that the technology had progressed to the point of eliminating the cordwood modules altogether and replacing them with all solid state electronics. The MTBF's were raised again to an incredible 300 hours.

Telephonics and Hughes used a technique called PCM (Pulse Code Modulation) in their multiplex systems. This technique is basically the conversion of the analog audio signal into a digital word. This digital word is typically a 10 to 13 digit code that represents all of the characteristics of the sampled slice of audio. This code is then transmitted throughout the system in place of the analog audio and then decoded and reconstructed at the receiver. Figure 8 (top left) shows a typical PCM waveform. The decoded signal is shown in Figure 8 (bottom left). Every 1/30,000 of a second each audio input is sampled, a slice is cut out, and fed into the multiplex stream. The demultiplexed audio signal does resemble the input signal somewhat but it is not a true reproduction.

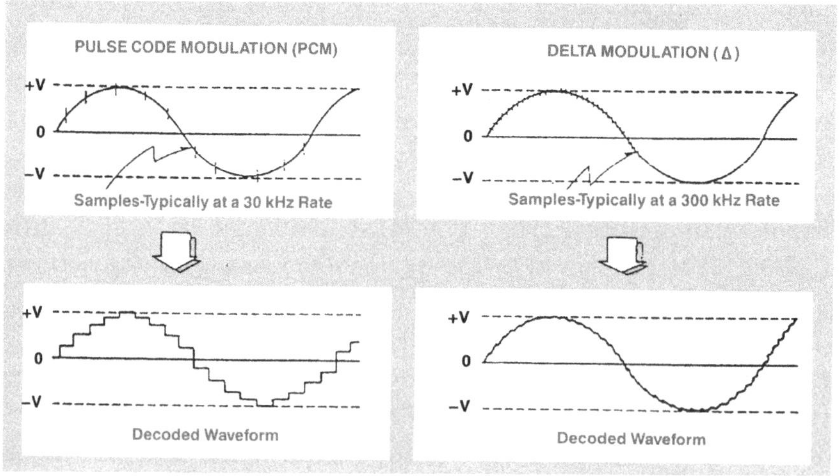

Figure 8 – Comparison of Delta Modulation vs Pulse Code Modulation

Matsushita Avionics Systems (MAS) entered into the MUX arena in 1980 with the introduction of its systems on the B767 aircraft. MAS developed a different type of sampling based on the Delta Modulation technique. This type allows for the sampling to be done about 10 times more often than PCM, or at about 300,000 times per second. Figure 8 shows a comparison of Delta modulation versus PCM.

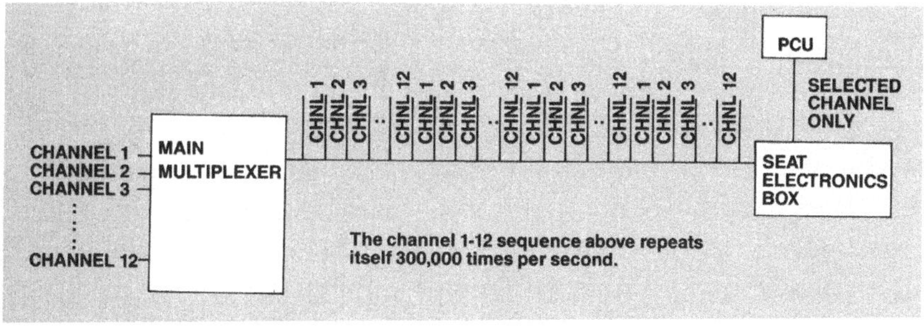

Figure 9 – Narrow-Body Aircraft Simplified MUX System

Adaptive Delta modulation samples only the change in the previous digital word (slice of audio), it doesn't copy the current signal. This allows for a greater sampling rate and therefore a truer reproduction of the input signal.

A typical multiplex system is represented in Figure 9. The multiplexed audio signals from each channel is fed down the wires one after another in a continuous flow. First a slice of channel one, then a slice of channel two, then channel three, etc. etc. This stream carries along with it a series of clock pulses that tell the receiver (Seat Electronics Box) what channel is where. Suppose a passenger selected channel six at the PCU. The PCU would transfer a signal to the Seat Electronics Box (SEB) to send it the audio program on channel six. The SEB would then monitor the Multiplexed data stream and listen for the clock pulses that said channel six was coming down the line. The SEB would then send the audio slice of channel six up to the PCU. This would happen 300,000 times per second. So fast that the human ear was incapable of discerning the separate slices.

To bring it all together, the audio is first input into the system by the Entertainment Tape Reproducer. This typically supplies 12 channels of audio to the Main Multiplexer on 24 wires, 2 wires per audio channel. Since the Main Multiplexer is typically located within one or two feet of the Tape Reproducer this poses no problem. The Main Multiplexer takes these 12 audio inputs and multiplexes them onto a single pair of wires or onto a coax cable.

In a typical narrow body aircraft application the Main Multiplexer has four outputs. Each output is fed to a different part of the aircraft and sent to the SEB's to be demultiplexed and then directed up to the PCU to be heard by the passenger.

Wide body aircraft have a slightly different scheme. Located within each zone of the main cabin is another box called a Zone Multiplexer or Sub-Multiplexer. This box is used to add the movie audio for that zone onto the multiplex stream along with the Tape Reproducer audio. This allows each zone in the aircraft to have a different movie playing, with the audio for each movie appearing on the same PCU channel in each zone. For example, if all movie audio is to be on channel one and two, then a passenger in zone one would have their soundtrack fed to them by their Zone Mux or Sub-Mux on channel one and two, and a passenger in zone four would have a different sound track fed to them by their Zone Mux or Sub-Mux on their channel one and two.

To trace the path more fully the Zone Mux/Sub-Mux sends its multiplex stream to the SEB's within its zone. There is one SEB per seat group. A seat group consists of one, two, three or four seats. There is one PCU per seat. The SEB is the demultiplexer and actually does the work of removing the audio channel selected by the passenger from the multiplex stream.

On wide body aircraft the reading and attendant call lights are also controlled by this same process. When a passenger presses the reading light button on his or her PCU (Passenger Control Unit) a signal is sent to the Seat Electronics Box that the light should be turned on.

The SEB (Seat Electronics Box) then waits until its turn in the multiplex stream to say if it wants anything done. When its turn comes it says turn on the light for seat "B". This is then decoded by other electronic devices and the light is turned on. This process causes the slight delay when turning on and off your reading light.

The control of the PSS (Passenger Service System) functions described above are accomplished by a single pair of wires using a similar multiplexing scheme as for the audio.

Troubleshooting this type of system used to be a very time consuming and complicated procedure. With the ability to design and manufacture microprocessors and LSI microchips, a system that can basically troubleshoot itself was developed. On a narrow body aircraft a BIT (Built in test) takes about 300 milli-seconds to run. That is about 1/3 of a second.

This system will test all audio channels at all volume levels on all Digital Passenger Control Units within the aircraft. The trouble spots are then indicated and in some cases what the fault actually is.

With the data lines that are available in a wide body aircraft multiplex system an even more comprehensive test can be run. The person conducting the test will know at a glance that the Seat Box in seat 31 ABC is bad, or that the cable assembly in the overhead between seats 12ABC and 13ABC has an open wire in its data #1 line, a bad PCU will be spot lighted by its reading light.

Although the ability to troubleshoot the systems has become more sophisticated, the actual need to do so has significantly been reduced. For example the MTBF of a typical Seat Electronics Box flying on a B747 is 310,000 flight hours, with an overall system MTBF in the 1,500 hour range.

So there you have it, a short history lesson and a not too technically detailed explanation of multiplex audio systems and how they operated.

Hybrid (Analog/Digital)

Hybrid systems took the basic Radio Frequency (RF) designs of analog Home Cable Television systems and ran them in parallel with a digital network for the Passenger Service System (PSS).

- Audio/Video
    - Transmitted over an RF modulated Network
    - Analog (Home Cable TV type) System
    - FM Based Proprietary Systems
- Passenger Service System (PSS) Commands
    - Reading Lights/Flight Attendant Call
    - Transmitted over Multiplexed Digital Network
- Wiring was in parallel and contained within the same cable bundle

PSS only on wide-body aircraft (Except in Business Jets which will be covered later).

These systems didn't fare as well as pure MUX systems, but they did "okay" in the market.

Digital Systems

- Ethernet: IEEE802.3
- Firewire/iLink: IEEE1394
- HD-SDI
- Proprietary/Other

Now we move into what would come to dominate the market, digital systems. On a side note, Wireless systems will be discussed later as they are a special case.

Basically it came down to two primary contenders, Ethernet and Firewire.

- Ethernet designed for office use: moving data
- Firewire designed for Home/Studio use: moving Audio/Video

The Sony Passport system was the first IFE system that incorporated Firewire technology and was revolutionary in its implementation. Unfortunately for it, their IFE division was purchased by Rockwell Collins who immediately killed that product line.

You see, it was a competitor to their own internally developed Ethernet based system. This would actually happen again years later when Rockwell Collins purchased Airshow and killed off their Firewire based IFE/CMS system for business jets, once again, in favor of their own Ethernet based system.

In any case, at the end of the day, digital systems all have one thing in common, and the name of the game is BANDWIDTH!

However, equally important in my mind is how you actually "use" that bandwidth. But I digress...

People talk about bandwidth all the time. But do you really know what they are talking about? Well, the more bandwidth you have then more data that you can have per second. The more data you can have per second the bigger and better video you can watch, or data you can transfer. But really, in most cases it's watching video that matters when we are talking about IFE.

Figure 10 shows the amount of bandwidth needed for different types of control, audio and video data. (Comp = Compressed data and UnComp = Uncompressed data)

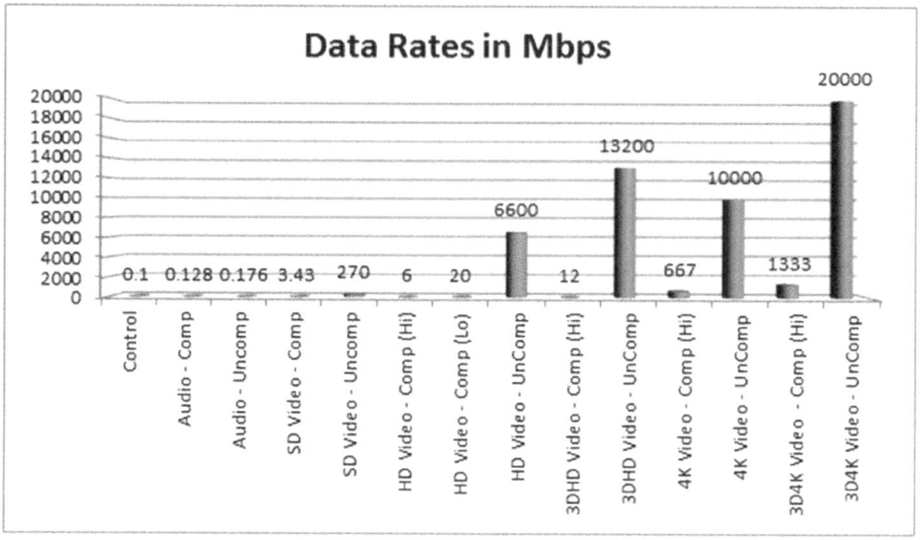

Figure 10 - Who needs what Bandwidth?

Also please note that the above chart currently only goes up to 4K Video. But Ultra-High Definition Television (UHDTV) (4320p) 8K video has been tested in commercial settings and is simply amazing.

So, how much bandwidth do you think you will need to have a dozen or so of those channels in an aircraft IFE system?

Lots! Lots and lots and lots and lots of bandwidth.

And when they move to 3D 8K, you can probably expect a bandwidth requirement of over 100Gbps per channel. That's PER CHANNEL! That's a HUGE amount of data.

If you're wondering why it's not in the chart, it's because it skews everything so badly the other data rates are mostly flat lines or little teeny bumps in the chart. Much like some of them are now, but even more so. And a bunch of little blips and one big tall line isn't very interesting. Yes, I could have made it using a logarithmic scale, but those are very hard for most people to read.

Ethernet (IEEE803.2)

Okay, back to current and past realities, Standard Definition (SD) and High-Definition (HD) video.

And of course back to that good old industrial complex, corporate office standard, Ethernet.

A few key points about Ethernet based systems:

- Typically Use Analog Source Equipment
    - Analog to Digital Converter Units

What this means, is that they take a regular old DVD player or video camera feed, run it through an Analog-to-Digital (A/D) converter, and then place this signal onto the data bus. A crude but effective method since the movie studios really don't like the idea of companies sending around unencrypted digital movie data.

- Industry Moving to all Digital Sources

Since many airline balked at having these extra A/D converter boxes, most of the suppliers started advertising "Direct Connect" source equipment.

Just hook the DVD player up to the network and away you go. How did they get the studios to agree to let them take the digital right off the disk and shuttle that around the cabin? They didn't.

The suppliers just took the electronics out of the A/D converter box, and cleverly repackaged it into the DVD players. So it was still the digital to analog back to digital conversion, but it was just hidden from view. And for all but the most discerning viewer or action packed movie, nobody could really tell the difference.

More common these days are Blu-Ray players with Component Video outputs that are converted to Digital as mentioned above, or with HDMI outputs that are encoded and placed onto the data network.

- Head-End File Server(s)

Here we have encrypted movies, directly from the studios, stored on massive hard disk arrays. Digital movies on demand. Brilliant for the airlines. The costs associated with handling all the tapes, the cost and logistics of the security they had to provide, all gone. Well, not “all” gone, but mostly all gone. And games... they could really add good games now. And newspapers, and magazines, and etc.

These units provided the same basic functionality as you would find in a file server on the ground.

- Zone Distribution Boxes/Data Concentrators

Here is where we moved to splitting up the data networks into zones in the aircraft. With each zone becoming its own little network. This was of course to reduce the amount of data traffic within each zone so as not to overwhelm the network. These boxes held various names depending on your company, country of origin, or just which sound you liked better when you said the name. No matter what you called it, it was essentially a data switch.

Eventually, and I mean many many years after they were introduced, these became more like smart routers, but many still flying today are just dumb Ethernet switches. (And I don't mean dumb in a bad way, "dumb" is really a technical term on how to describe them.)

✈ Seat Electronics Boxes (One per Seat Group)

Remember my description of MUX system earlier and the seat box? Same thing here except it's a smarter box.

They usually come in three or four seat versions. Meaning they can serve three or four clients at one time.

The seat boxes contain all the brains and the monitors and PCUs in the seats are just dumb terminals. The seat box acts as three or four micro-mini Personal Computers. With the PCUs and handsets acting as keyboard/mouse controls and the in-seat monitor acting as, well, an in-seat monitor.

However, there are some systems that, in an effort to increase the life span of their existing designs, adopted the use of smart monitors along with their existing, and slightly modified, seat boxes. The smart seat-back or in-seat monitors really taking concepts of laptop computers and adding touch-screens. These were essentially "built-in" tablet PCs, before there were "portable" tablet PCs.

✈ No-Seat Electronics Boxes

There are also systems with no seat boxes. Each seat is basically wired monitor to monitor, with a small junction box to route and distribute power to them. These systems almost never have a PCU associated with them; control is nearly always done on just the touch-screen.

✈ Analog/Digital Peripherals

These are all the bits that hang off of the seat boxes or are connected to the network as clients.

These take in the digital data, and in most cases, convert it to analog for presentation to the passenger.

We'll look at these one at a time:

1) Monitors

Come in two flavors:

- Bulkhead
- In-Seat

Bulkhead monitors are affixed to the walls, or ceilings, and are typically, but not always, direct connect digital devices. It really depends on the monitor sizes and where they expect to be mounted and the technologies involved. Generally speaking, the larger the monitor, the more likely it is not to be directly connected to an Ethernet data network.

In-Seat monitors are about 50%/50% smart and dumb. Some are directly connected to the Ethernet network and some aren't.

2) Passenger Control Units (PCUs)

These have the controls for the audio channel selection, the video channel selection, the volume control, reading lights, and Flight Attendant Call, Etc.

These can also be full-up handsets with QWERTY keyboards, telephone functionality, credit-card readers, etc. Historically these used either proprietary data interfaces or RS485. Mostly today they are almost all USB connected devices.

Most of the time, smart companies will provide audio to the passenger headset jack by routing it from the PCU (or smart monitor). This is because if they do the digital to analog conversion in the PCU, there is the least likelihood of noise and distortion of the audio signal to the passenger headset. However, some companies still do the decoding in the seat-box and then route long, unshielded, cables to the headset jacks. Not the best way to do it, in my humble opinion.

3) And last but not least, passenger supplied devices

Most likely these are laptops, netbooks, tablets or other similar devices that will access an on-board to off-board data network via Wi-Fi for e-mail and web browsing.

Figure 11 shows a simplified block diagram of a basic star wired Ethernet system. Again, this only shows a small portion of a cabin, it would be a nightmare if an entire B747 was wired like this. But maybe for a B737 or A320 it could be done, still, it wouldn't be pretty.

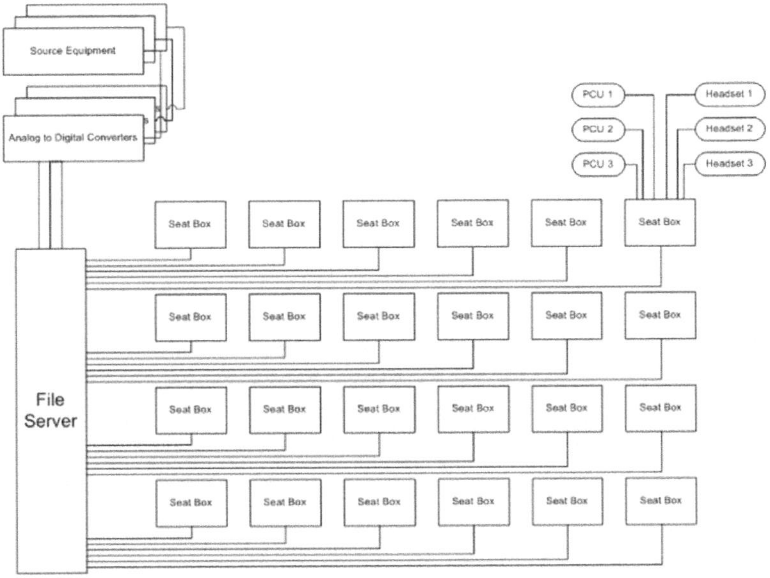

Figure 11 - Ethernet Block Diagram (Star-Wired)

Figure 12 shows two different types of systems, both based on daisy-chain methodology.

On top is daisy-chain wired from a zone box, while on the bottom is star wired from the zone box. Please note that an installation would not typically consist of both types in a single aircraft installation. I'm showing them together here because it's easier to show the differences, and similarities, when they are so close together in the same drawing.

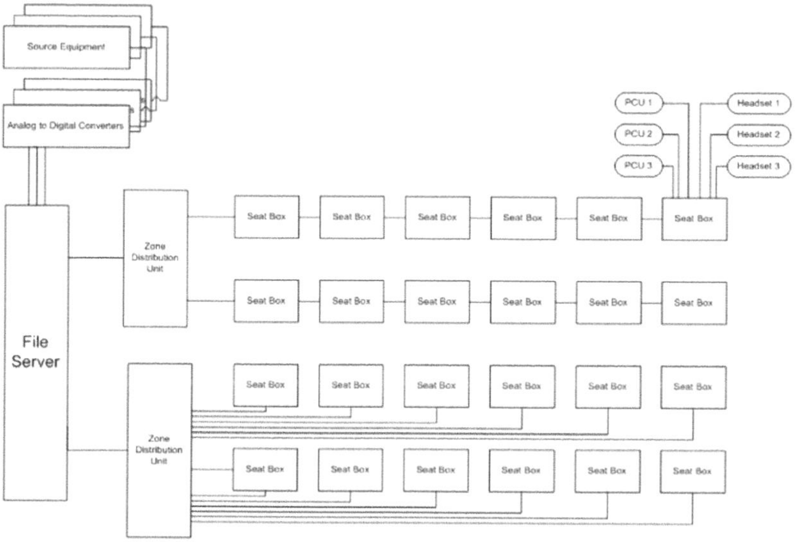

Figure 12 - Ethernet Block Diagram (Daisy-Chained & Zone-Star)

So, the beginning, or head-end of both types is the same, the file servers connect to source equipment, and sends that data out into the cabin to a zone box, or zone boxes, within each zone. Then from the zone boxes out to the seats is where it gets different. Either being routed daisy-chain fashion from seat box to seat box, or star wired out to individual seat boxes.

Ethernet also has a little draw-back in its implementation when used as a medium for audio and video. That is, it is based on a CSMA/CD access method. CSMA/CD meaning Carrier Sense Multiple Access with Collision Detection.

Which is great in an office environment with emails, print jobs, web pages and data files being sent to and fro. However, for audio and video, which counts on an uninterrupted and steady stream of data, it is very problematic. And the issue grows worse and worse as the number of channels increases.

So, why is CSMA/CD important?

Well, because it determines who gets to talk, or transmit, on the network. Say you have 100 nodes on the network. They are all just sitting there listening. One of them wants to talk/transmit, so it listens (CARRIER SENSE), if the line is not being used it transmits.

Now suppose node 1 is transmitting and node 50 wants to transmit, node 50 waits until node 1 is done and then it can transmit. But further suppose that nodes 25 and 26 were also waiting to transmit. Now, as soon as node 1 finishes then nodes 25, 26 & 50 all transmit at the same time. A collision takes place on the network.

This brings us to "Truncated Binary Exponential Backoff".

After the network detects a collision, the nodes that wanted to transmit all have a random reset timer activate and then try again. The reset timer though has rules:

- After the 1st collision the senders wait 0-1 time slots then try again
- After the 2nd collision the senders wait 0-3 time slots then try again
- After the 3rd collision the senders wait 0-7 time slots then try again, etc., etc., etc.
- The Default max wait value is 1023 time slots
- Wait time is 2C-1 (c = number of collisions)
- One time slot is typically 51.2μS

What this means is that transmissions are interrupted and you could end up in a situation where one node "never" gets an opportunity to transmit. Here is an example of "Ethernet Capture":

1) Node 25 and Node 26 both transmit and collide
2) They both randomly pick a wait interval. Node 25 wins and transmits, node 26 waits
3) Node 25 finishes sending that packet and node 25 & 26 try and transmit again, and collide
4) Node 25 waits and since Node 26 has a larger wait pool to select from, because it lost last time, so it is likely to lose again
5) Next packet time to send, node 25 and 26 collide again
6) Node 25 has a smaller pool of time slots so it wins again
7) Node 26 pretty much will never get a chance to transmit as long as Node 25 has anything to place on the network, like a two hour HD Video stream.

Now imagine you have 20 or 30 of these HD video streams to put on the network. Big problem huh?

This was really an issue with 100Mbps Ethernet systems, but not so much with 1,000Mbps systems, which I will get to in a minute.

Some of you that know Ethernet may be saying, "but what about QoS and CoS improvements? QoS being Quality of Service and CoS being Class of Service.

These absolutely improve the ability of Ethernet to handle video in a home or office, but if "all" the traffic on the network is video, and "all" of it is of the same class and quality level, it doesn't really help. If anything, the overhead of QoS and CoS greatly reduces the theoretical maximum bandwidth you have to play with which exasperates the problem even more. If "ALL" of the traffic is "High-Priority", than effectively none of the traffic is.

Moving to Gigabit Ethernet, it did do one thing with CSMA/CD, it increased the possible wait time slot from 512 bits to 512 Bytes (4,096 bits) so if you are caught in a collision, you have a much longer wait potential before giving up.

Also, if your Gigabit Ethernet connection is operating in full-duplex mode, there is no chance of having a collision. But full-duplex is only available for point-to-point connections, not through a hub that connects several devices.

But even then, you can have the receiving MAC (Media Access Control) buffer fill up and have to send a command to the transmitting node to stop transmitting, so that you can clear the receive buffer.

So, while almost every IFE supplier and their brother have adopted an Ethernet topology, there are still issues that remain. The solution so far has been to try and install faster and faster networks, or, more and more networks and sub-networks, or smaller and smaller network zones, so there are fewer nodes that can be collided with on each individual network or sub-network. Another tactic is to use higher and higher levels of data compression to squeeze the video into a smaller and smaller set of bits.

Now some of you may be thinking, "wow, he really hates Ethernet". I really don't. Actually, I have seen it implemented very well and quite effectively. Unfortunately, I have also seen Ethernet systems that were developed horribly and didn't turn out so well. It all depends on the system architecture, the software, the purpose and the implementation.

Digital/Other

Besides Ethernet, there are several other proprietary system implementations. For the most part they are all variations on the multiplexing methods I described earlier. (See, I told you it would be best to read this book in order.)

Some of these "other" IFE systems are stand-alone networked devices. There are really and endless array of possibilities here.

I have a much larger section later on Fiber Optics but I thought I'd mention it here.

While used sporadically over the years, Fiber Optic IFE systems haven't gained in popularity as much as copper systems have stayed dominant.

There are a lot of reasons for this, one being that companies that developed Fiber based systems have routinely been purchased by the big guys and then their development and systems were shut down.

Also, the extremely high cost of connectors has been one barrier, maintenance has been another. However, the connector costs have come down significantly, but the perceived maintenance issues remain.

I remember talking with a mechanic at an airline once about fiber optic systems and he said, "Yeah, it sounds good, but how do you measure 5 volts of light?" So, there is still a lot of training that needs to take place before airlines and Mechanics are comfortable with Fiber in the cabin.

A few years ago a company came out with a fantastic Fiber Optic based system that is gaining traction in the industry and I'll talk about it in more detail shortly.

More Modern Implementations

All digital still of course but with a bit of a twist, systems that don't have seat boxes at all.

Figure 13 shows a typical installation that doesn't use seat boxes. Well, "technically" there is still a small junction box for power and to provide a disconnect for the data cabling, but it doesn't have electronics in it so it's not considered a "Seat-Box".

In these systems, the seat-back monitor or in-arm monitor, is a "smart" monitor. This means that the intelligence is built into the monitor, not into a seat box. Basically, these monitors are like a laptop computer without a keyboard. I'd say like a tablet before there were tablets but these are much more powerful, and not battery operated.

These smart monitors are typically running some flavor of Linux, or in some cases embedded Windows.

So far they have received the most traction in Regional Jets and in Single-Aisle/Narrow-Body aircraft.

The data cabling is shown to go directly to the smart monitors, but "typically" it is routed through the junction box so that the cabling is made easier. It was just visually easier to show the different cabling by separating them out.

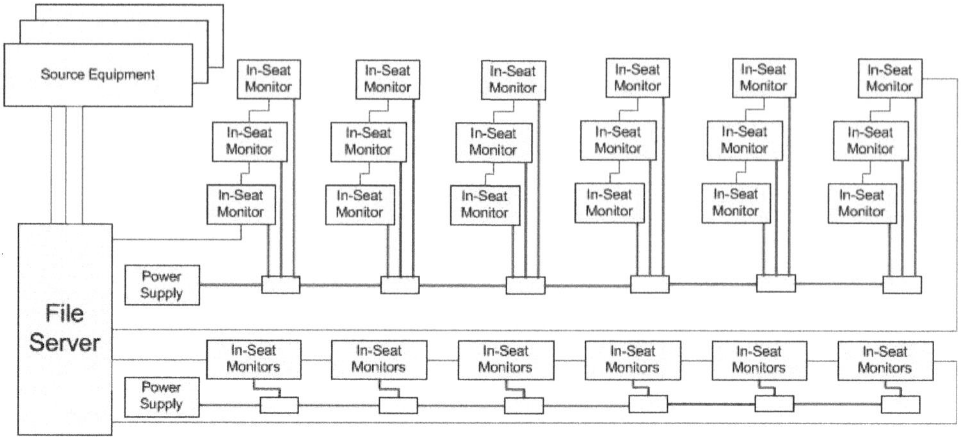

Figure 13 - Typical Digital Block Diagram

Coming up next in Figure 14 you will see what happens when we add "In-Seat Power" to the mix.

Here comes the seat box. Why? Because it's much too difficult to shuttle around all that DC power for laptops in the seat unless you have local power converters.

Keep in mind that aircraft power is traditionally 115Vac 400Hz, and now with wild frequency systems it's 115Vac 360Hz to 800Hz. Laptop computer chargers, cell phone chargers, tablet chargers, etc., would not be too happy of you tried to feed them that type of voltage & frequency.

So, on the data side adding in-seat power to the mix is identical to the non-in-seat power system. But on the power side, we add the power converters at each seat group. These then supply power for both the smart monitors and to the in-seat power ports.

The data cabling is shown to go directly to the smart monitors, but "typically" it is routed through the seat box so that the cabling is made easier.

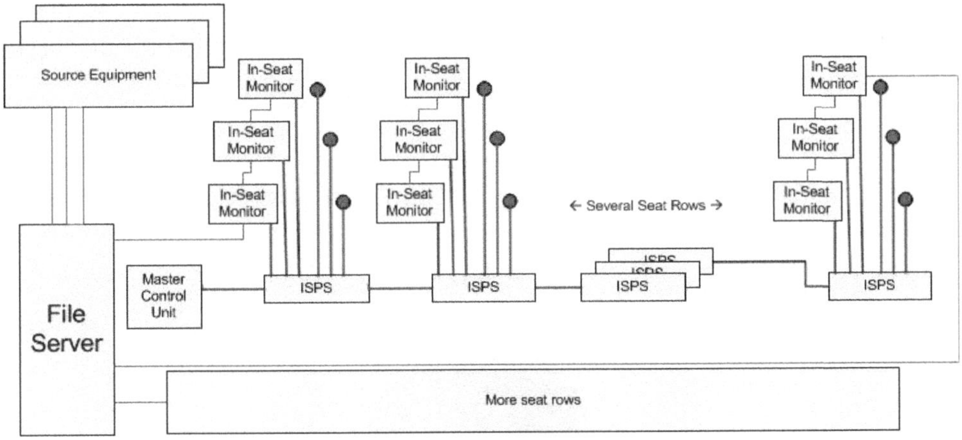

Figure 14 - Typical Digital Block Diagram with In-Seat Power

Fiber Optic Systems

Now, on to Fiber Optic based systems.

These systems are "still" trying to gain acceptance. I think the first Fiber based system I saw that was purpose built for an aircraft IFE system was back in 1991.

These installations are typically a combination of Star and Daisy-Chain. With the fiber optic and power cabling going from the head-end, or multiple head-ends, to the smart monitors and obviating the need for seat-boxes.

These systems are typically hundreds, if not thousands, of pounds lighter than their Ethernet counterparts. (Depending of course on the size of the aircraft, implementation, etc.)

I mentioned before about the perceived maintenance issues with Fiber Optics, which still holds true. But the mechanics are getting better training for these new IFE systems, as many cockpit (Flight Deck) avionics systems are moving to Fly-by-Light implementations. So a side benefit is that the

IFE use of Fiber is getting more acceptance in the maintenance community.

The one issue though that still needs to be kept in mind, is that the use of fiber optic cables means that the cable terminations need to be precise and that the connectors, when not connected, need to be kept immaculately clean.

Loops/Columns per System

As I mentioned before, in the world of IFE, the name of the game is BANDWIDTH.

In looking at a typical Fiber Optic based system implementation we can have the following:

- Systems with as few as 18 seats per string and as many as 45 seats per string
    - These are daisy-chain nodes
- For an average 21 Seat column:
    - Fiber optic systems typically provide 80Mbps per seat
    - Ethernet systems typically provide 3.8Mbps per seat (at 100Mbps Ethernet)
    - For a Gigabit Ethernet it's about 40Mbps per seat...

So, how many more video channels can I provide on an 80Mbps link than on a 3.8Mbps link? Or a 40Mbs link?

Video on a commercial air transport aircraft is usually highly compressed; many people can easily see the compression artifacts and find it distracting. With more bandwidth, you can have less squeezing of the video signal, and thereby provide a much truer video presentation. Which many in the commercial airlines may not care about, but what about the VVIP privately owned B747 with 108", or larger, Super HD (8K) monitors installed?

Wireless IFE (Embedded)

Ah the mystique and allure of "wireless" systems. Just put boxes anywhere and have them talk wirelessly, no more wires to run... Think of all the weight savings if we don't have to run wires anymore!

Really? Think it's that simple? We're not talking about wireless data to passenger carry-on devices; we are talking about embedded IFE electronics boxes in the aircraft, communicating with each other via wireless RF transmissions.

Just keep in mind that you still need to route "power" wires to those boxes! And now you have to install many wireless routers around the aircraft and wire up power to them as well.

Wireless was getting a lot of press and a lot of attention in the aviation marketplace over the past few years. Even Boeing had initially mandated that the B787 had to have a wireless embedded IFE system. Until they got the designs and some systems to test with. Turns out the "wireless" system was a lot heavier than a traditional wired system, and a lot less reliable than a wired system as well. So they scrapped that idea and went back to a simple wired system.

Seat vendors also complained rather loudly about implementing wireless, in that the in-seat wiring was now much more complex. They had to run wires (antennas) in the seat and generally these seats ended up with the most complex cabling.

With regard to wireless capabilities, as I'm writing this, IEEE802.11 dominates the global wireless market.

IEEE802.11b/g/n pretty much dominates the space and 802.11ac picking up speed as the new "standard". Although you need a pretty robust set of radios and multiple antennas to take full advantage of the specifications full capabilities.

	Theoretical	Real-World Best Case	Typical User	Proprietary Theoretical	Typical Proprietary
802.11b	11Mbps	4-5Mbps	2-3Mbps	22Mbps	6-7Mbps
802.11g	54Mbps 36Mbps 24Mbps	15-24Mbps	10-15Mbps	108Mbps	N/A
802.11n	600Mbps 450Mbps 300Mbps 150Mbps 65Mbps	Varies significantly Up to 400Mbps at top speed.	About 60% of the rated B/W for that connection.	N/A	N/A

Figure 15 - IEEE802.11b/g/n Throughput

And below is a special chart just for 802.11ac.

Theoretical Aggregate	Theoretical MAX to any one device	Typical Target Client	Antenna Configurations
6.77 Gbps	1.69 Gbps	Digital TV, Tablet, Laptop, PC	8-antenna AP four 2-antenna STAs 160Mhz (MU-MIMO)
6.77 Gbps	3.39Gbps to 4 ant STA 1.69Gbps to 2 ant STA 867Mbpa to 1 ant STA	Digital TV, Set-top Box, Tablet, Laptop, PC, handheld	8-antenna AP - 160 Mhz (MU-MIMO) one 4-antenna STA one 2-antenna STA two 1-antenna STAs
3.39 Gbps	867 Mbps	Handheld	4-antenna AP - 160 Mhz (MU-MIMO) four 1-antenna STAs
1.69 Gbps	1.69 Mbps	Tablet, Laptop	2-antenna AP - 160 Mhz 2-antenna STAs
867 Mbps	867 Mbps	Handheld	1-antenna AP - 160 Mhz 1-antenna STA
867 Mbps	867 Mbps	Tablet, Laptop	2-antenna AP - 800 Mhz 2-antenna STAs
433 Mbps	433 Mbps	Handheld	1-antenna AP - 80 Mhz 1-antenna STA

Figure 16 - IEEE802.11ac Throughput and Antenna Configurations

But, you still have to install wireless access points, wire them up, etc.

Wireless IFE (To Passenger PEDs)

Wireless IFE to passenger devices has really taken off the past few years. While the wireless promise of an embedded system has not borne fruit, the wireless transmission of IFE to passenger tablets, cell phones and laptops is gaining ground nearly every day.

This is where the airline stores content on-board a server or two, and then streams it to a passenger device via the on-board Wi-Fi.

Some systems require that the passenger load a special application prior to boarding, others let you connect with any web-browser. Some of the more interesting implementations are as follows.

- Qantas
  - Commissioned books that can be read and completed in a single flight
  - They provide IFE wirelessly to an "airline provided" tablet.
  - They also provide some A/V entertainment to passenger carry-on devices.
- American Airlines
  - Audio and Video entertainment to passenger devices
- Delta Airlines
  - Options for Audio, Video, Flight Information & Social Networking
- Air China
  - Options for Audio, Video, Games & Social Networking
- Alaska Airlines
  - Moving away from the portable digEplayer to the GoGo Vision system and providing IFE to Passenger Devices instead of having passengers renting a device from the airline
- Condor
  - Audio and Video options
- Virgin America
  - Audio Video selections added to existing embedded IFE offering
- Virgin Australia
  - Options for Audio, Video & Games
- GOL
  - News, Magazines, Audio/Video, Games, etc.
- AirAsiX
  - An eclectic mix of content, some free and some chargeable
- United Airlines
  - Adding Wireless-IFE to enhance their embedded system offerings.

- ✈ Southwest Airlines
    - Currently has Live and on-Demand IFE to PEDs via the Dish Network
    - Passengers can borrow an iPad from the airline if they don't have their own

What most airlines are trying to get to here is to provide passengers with entertainment options but without the high cost of installing an embedded system. Although some are using it to augment their embedded systems, they are really in the minority.

The biggest hurdle here however is still the reluctance of the movie studios to provide "Early-Window" releases to a "Passenger Owned" device. There are certainly carriers that that have Early-Window releases on Airline Supplied tablets, but not to passenger devices.

Right now it is just too easy for a technically savvy passenger to copy pretty much any video that they have access to on a laptop screen. And until that issue is solved, it's unlikely that the studios will agree to passengers having access to their most lucrative movies.

Some airlines have gotten around this somewhat by handing out portable devices in First Class with premium content embedded in them. And have the units use their wireless link for the non-premium content being distributed to the passenger devices. I chat more about this later.

Wireless IFE (Jamming)

Many aircraft today use microwave ovens. This is especially true in Business Jet and VIP/VVIP aircraft.

What you may not be away of is that typical microwave ovens generate their microwaves using a device called a magnetron. And because of the allocated frequency bands that the microwave ovens can operate in, you will find them all operating around 2.4GHz. Which just so happens to be the same frequency used by most Wi-Fi systems.

Microwave ovens in the home usually run on a half-cycle transmitter, while those used in industrial settings like restaurants operate on full-cycle transmitters. (Actually they use two half-cycle transmitters.)

In nearly all cases, when the magnetron is generating microwaves (i.e. transmitting) there is some Radio Frequency (RF) leakage from the unit into the surrounding area. This results in the jamming of things that use similar frequencies, such as Wi-Fi networks.

Many aircraft OEMs actually have restrictions in place as to how closely microwave ovens can be installed to each other.

Figure 17 shows the waveform of a typical unit found in a home and in some aircraft.

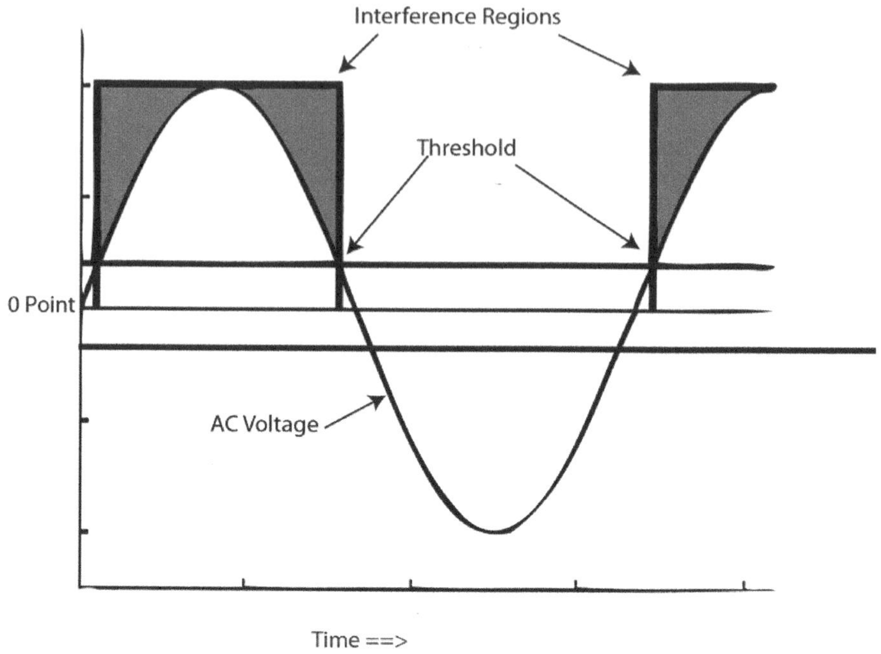

Figure 17 – Half-Cycle Microwave Transmit

The units are designed to use the positive transition of the input AC power to turn the magnetron on and off. So, for every ½ cycle of the input power's frequency, the unit is effectively jamming nearby systems in small doses for effectively half the time that the device is operating. This jamming reduces, but does not totally prevent, the operation of nearby Wi-Fi and many cordless phones.

This will typically reduce the available Wi-Fi bandwidth to about half of what was available with the oven turned off.

Figure 18 shows the waveform of a typical unit found in an industrial setting like a restaurant, and also can be found in many aircraft.

In order to cook faster, these units are fitted with two magnetrons, one that transmits during the positive cycle and one that transmits during the negative cycle.

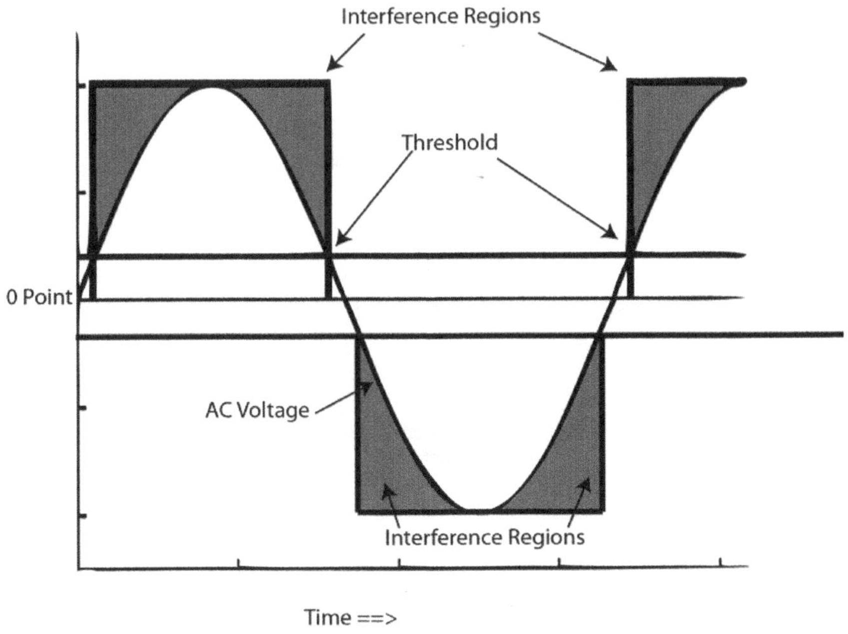

Figure 18 – Full-Cycle Microwave Transmit

So, for every ½ cycle of the input power's frequency, either one or the other magnetron is transmitting, which effectively jams nearby systems for the entire time that the device is operating. Except for that small smidgen of time as the power crosses over a small band near the zero point.

This jamming prevents the effective operation of nearby Wi-Fi and other devices operating in the 2.4 GHz band. Which brings the available Wi-Fi bandwidth to zero, or so near zero that you basically have no bandwidth.

And yes, there are several ways to mitigate and eliminate this RF interference, but that is beyond the scope of this tome.

Oh, and for the DC powered ovens, they use an oscillator to simulate the on/off triggers of the typical AC power input.

General IFE System Functionality

Alright, now you should have a good idea about what the hardware behind the systems is about. So, what do you do with all that hardware?

Let's start with something simple.

Broadcast Audio & Video

In this category:

- All passengers receive the audio or video program at the same time.
- Passengers have no transport mechanism controls (e.g. Play, Stop, FF, REW, etc.).
- Same principal has home radio or television broadcasts. (Television without TiVo or DVR capabilities.)

That is not to say that the passengers only have one channel, quite the contrary, they may have 20 or 30 or more channels to choose from. They just can't pause, rewind, fast forward, or have any transport controls. They simply change channels and watch what's playing on that particular channel at that particular moment in time.

The passenger just tunes in (channel select) what type of programming they want to have. The key here is that passengers have no control over the program start/stop or transport controls. If they tune in during the middle of a program, then that's where they begin...

This is considered to be the BASIC IFE offering from any vendor. Audio "only" systems are "bare-bones" basic.

Now, before I get too far I'd like to mention a "Rule of Thumb" for video monitor placement.

- 1 inch (2.54cm) diagonal measurement of the monitor = 1 foot (30.48cm) viewing distance

Of course you can see a monitor from further away, but this is a rule of thumb for optimum top end viewing distance. For every 1 inch diagonal size of the monitor, you can see the monitor and make out what is going on about that many feet away.

Satellite Television (Conventional)

For a conventional satellite TV system we have:

- An antenna that is used on the outside of the aircraft that can receive satellite TV signals
- The antenna can be either a dish type (single or multiple feed-horns) or a phased array antenna

- This is typically a "Broadcast" video application
- Multiple channels are available as are video decoders for multiple regions
    - (E.g. USA, Middle East, Europe)
        - (E.g. Middle East = Arab Sat & Nile Sat)

The one really key point to remember here is that if there is no satellite broadcasting to the ground in a particular area, then there is no coverage in the air either... Antarctica for example... Or in the middle of the Atlantic Ocean.

Nobody is going to waste satellite air time to broadcast video to fish. You're only going to find coverage over land or very close to land as the signal spills out a little bit over the water.

Also special decoders are needed for various regions of the world. A decoder designed to operate in US airspace will not receive European signals when flying over Europe. There are also issues with satellite signal polarization Horizontal versus Vertical. Different regions have different requirements and different services providers within each region have different requirements.

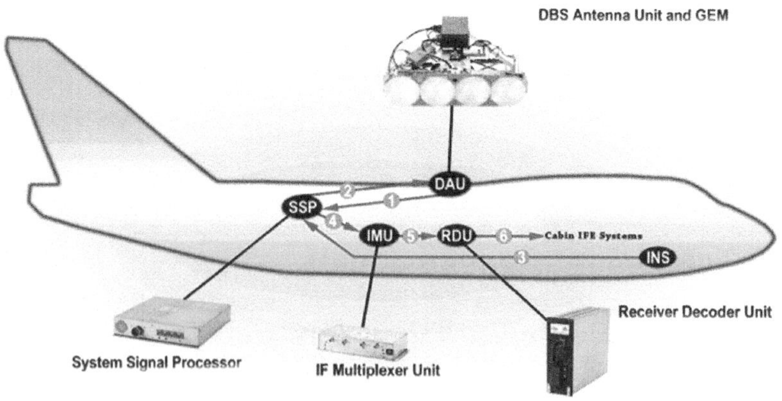

Figure 19 - TW550 Satellite TV System

The raydome, which covers the antenna, is a critical part of the TV system that the IFE suppliers typically do not make.

Figure 19 shows the Rockwell Collins TW550 system for large Air Transport type aircraft. The RDU in this case has slots for two received cards so you get two video channels per RDU.

Rockwell Collins also makes tail/fin/vertical stabilizer mounted systems for smaller aircraft like the Gulfstream G550 or the Embraer Legacy and similar such aircraft.

Please also note that you could have many RDU's installed in this system. You are certainly not limited to only having two (2) TV channels. One installation that I was a part of had nine (9) RDU's installed.

Satellite Television (via Internet)

Here is another area where it gets interesting. Remember earlier I mentioned that Connexion by Boeing had delivered a TV over the broadband Internet solution? Well, Panasonic Avionics is carrying the torch with a similar methodology now having purchased many of the CBB assets when CBB closed shop.

This solution offers multiple satellite TV channels carried over the satellites' broadband internet data network. As with the traditional Satellite TV Signal, this type of video is also "broadcast" video and not interactive.

The one advantage these systems have though is that they have nearly global coverage. Near as I can recall, Antarctica is still out of the coverage area as is the North Pole. But outside of that, the coverage is enormous when compared to traditional Sat TV systems.

However, the number of channels they have available is miniscule when compared to the numbers that are available with traditional Sat TV.

So pretty much the tradeoff is a lot of channels in a medium size area, or a few channels in a very large area.

Although, I am aware of several instances where both types of systems have been installed onto a single aircraft. These have been in the private aircraft world, not in commercial air transport.

Satellite Radio

About the only player in this space has been Heads-Up Technologies which has/had the exclusive license to offer the XM Radio network to aircraft flying in the United States.

Sirius Radio never appeared to show much interest in the aerospace market and so Heads Up, with the XM Radio service, held 100% Market Share. Oh, and in case you weren't aware, in 2007 XM Radio and Sirius merged to form SiriusXM Satellite Radio.

Right now if we are talking strictly about Satellite Radio, in the Continental US (CONUS), SiriusXM is about the only game in town.

And for my Canadian readers, their parent company has merged the Canadian versions into SiriusXM Canada. Although I haven't heard of any aircraft fitted with systems operating in Canadian airspace, there very well may be. In which case, please contact me and let me know who is operating it and where. Thanks!

Anyway, back to the narrative, these installations are nearly identical to the Satellite TV installs except on a much smaller scale. Smaller antennas, smaller down converters, smaller amplifiers, etc.

While these are frequently fitted as stand-alone systems, they are occasionally offered as part of a larger IFE system retrofit by subcontracting from one of the big IFE companies.

- An antenna is used on the outside of aircraft that can receive satellite Radio signals
- So far only available in the Continental United States
- Multiple channel radio receivers are available
- Stand-Alone or as part of the IFE system

Moving Map

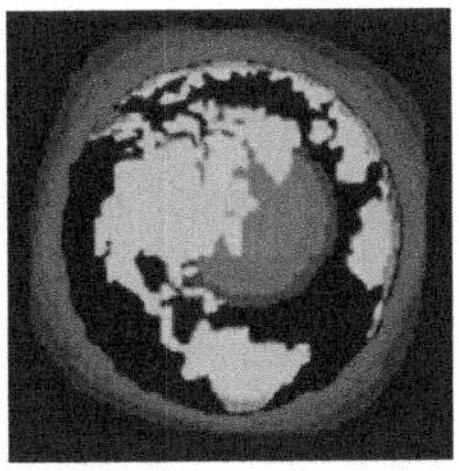

If you've flown in a commercial aircraft within the past twenty years or so the odds are pretty good you have seen this product.

It provides a visual indication of where in the world you are. Usually an icon or 3D model of an aircraft is positioned in the middle of the video screen and the terrain passes by underneath.

The image of the terrain could be a zoomed out globe of the Earth to a zoomed in view of city streets, or anything in between.

Since these systems are connected to the avionics systems data busses, you also can receive and display Flight Parameters such as:

- Altitude
- Air Speed
- Ground Speed
- Time to Destination
- Outside Air Temperature
- And many many other parameters

I recall one instance where a Business Jet owner wanted the aircraft Fuel Quantity shown on his Moving Map. I can only guess what incident caused him to want that parameter displayed...

And speaking of displaying odd things, how about Outside Air Temperature? Every Moving Map I have ever seen has shown this parameter as basic to their passenger display. Why?

Okay, so we know it is -57°C outside the aircraft while in flight. Is anyone going to go out there? Have a scenic stroll around the wings? It just seems like a waste of screen space to me.

Which brings me to zooming in to street level. Things here can get pretty cluttered with the ground zipping along pretty fast. So these types of zoomed in views are typically restricted to take-off and landing. Where you're flying into or out of a city and want to show a better perspective.

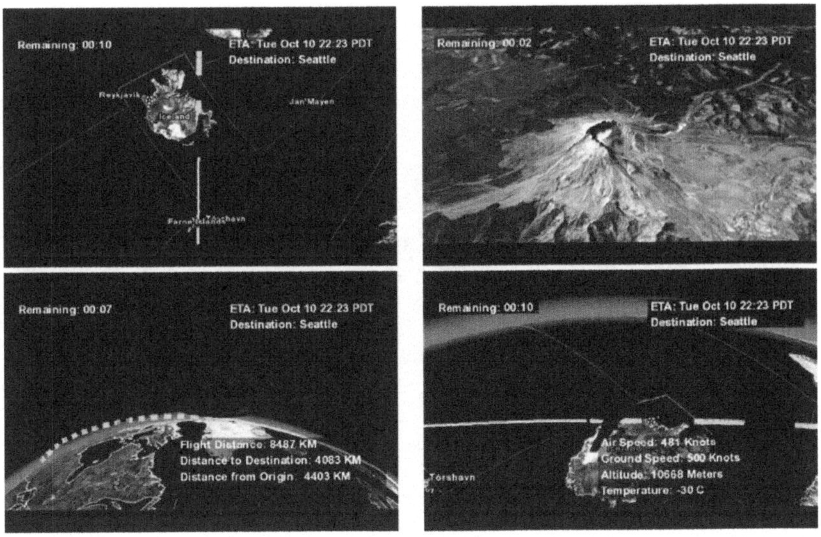

Figure 20 – Typical Moving Map

Oh, and in case you are curious, the image in the upper right is of Mt. St. Helens in Washington State, well after it erupted. The whole top and part of the side of the mountain blew off in 1980.

Connecting Gate and Arriving Gate Information

This is information that is displayed on the cabin video screens that basically replicates what you would see in an airport terminal. It is usually provided as an option to the Moving Map package.

The really great feature here is that you will know that if you are late arriving for a connecting flight, if your outbound flight is also delayed and you can leisurely make your way to the next gate. Or, if you have to sprint like a gazelle through the airport to make the connection. Or, if you've missed it completely and have to slog your way to the customer service counter.

Figure 21 - Connecting Gate Information

Real-Time News and Weather

This is a feature that isn't nearly as widespread as it was expected to be, at least not in the Commercial Air Transport world. It seems to have a lot more of a following in the Business Jet and Private Jet sectors.

This service provides a variety of real-time information such as:

- Stock Quotes
- Sports Scores
- News Bites

Like the connecting gate info, this is a service that is usually an option to the Moving Map offering.

Essentially the service operator contracts with news agencies to allow them to reformat web pages and newsfeeds and then broadcast them in formats usable to the aircraft.

In the Air Transport world, "censored" News and Information is very common. Typically airlines will provide the service operator with a list of keywords that if they appear anywhere in the story then the story is then not sent to the aircraft.

Some typical restricted words include:

- Hijack
- Terrorist
- Crash
- Bomb
- Deep Vein Thrombosis
- Explosion

And there are hundreds more.

Boarding Music

The relaxing music provided to soothe you while you board and exit the aircraft.

Pre-Recorded Audio and/or Video Announcements

More and more airlines are using these automated announcements so as to reduce the work load on the Flight Attendants, and to have a more consistent passenger experience.

These messages typically consist of:

- Welcome Aboard
- Safety Briefings
- Destination Information
- Customs and Immigration Procedures and Forms
- Any similar type message or briefing

When used with the on-board video systems, it also relieves the Flight Attendants from doing the manual demonstrations of how to fasten your seat belt, don the oxygen mask, find the nearest exit, etc.

Some of these safety briefings are animated cartoons, some use live actors and some just use still graphics. As long as the safety message gets conveyed, the FAA, CAA and other global equivalents don't seem to mind what form the messages are in, just so long as they are understandable.

PRAM/BGM

PRAM/BGM is the shorthand we use when talking about Pre-Recorded Announcement Machines and Boarding Music Machines. Because in the beginning, these were in fact separate units.

Then over time they became one unit. Many aircraft are still flying with them today. On many narrow-body (Single-Aisle) aircraft you can find them on the wall above the Flight Attendant's seat just as you enter the aircraft.

However, for the most part, today they are typically just software applications running on a server. Certainly on Wide-Body aircraft, this function is typically provided via parts of the IFE equipment:

- Fed from a spinning Hard Drive in a File Server
- Fed from Solid State Memory
- CD, DVD or Blu-ray
- Audio tape or video tape is still used in some parts of the world but pretty much being replaced by digital storage

In addition to be included in some IFE systems, I believe all of the Moving Map vendors also offer these features in their products as well.

Audio/Video On-Demand (AVOD)

AVOD, for most, is the pinnacle of the IFE world. Passengers can watch what they want, when they want it. The key difference between AVOD and Broadcast Video is that with AVOD the passengers have their own "Transport Controls".

With their own transport controls passengers can play, pause/stop, rewind & fast forward through a movie to their hearts content.

Every passenger gets their own, personal, video stream to do with what they will. Want to see that car explode again? Rewind and play it again. Want to skip through the boring bits of a movie? Hit the Fast Forward button.

The source of the AVOD content is typically from sources like these:

- Streamed
    - Head-end File Server(s)
    - Stand-Alone Hard Drive Array attached to the network
        - Network Attached Storage (NAS)
    - "Distributed" head-end functionality
- Local (fixed) at each seat
    - DVD/Blu-ray Video Player in the seat
    - Local hard drives/PCs/Tablets affixed to the seat
- Device handed out by Flight Attendant
    - Special purpose video players like the digEplayer™
    - Other: Portable DVD Player's, Tablet PC's, etc.

A couple of quick notes on the above:

1) The "distributed head-end" functionality was originally pioneered by Henry Debey at Delta-Beta. He had a great concept of moving all the head-end content and services out into the cabin and "distributing" it to run from the seats. This made his systems extremely fault tolerant and the proprietary video compression algorithms he developed and used made very efficient use of the limited network bandwidth available at that time. Perhaps his system was too far ahead of its time as it only saw limited success in the marketplace.
2) There was also a time when there were Hi-8 video players build into the seats, mostly in First Class, where the Flight Attendant would come by with a box of tapes and you would pick the one you wanted, put it into your player in the seats, and watch away. Oh the horror stories I could tell you about those things...

The other major thing that AVOD brought to the airlines is the complexity of the new content licensing agreements. I'll cover that later in the section on Programming and Content.

Passenger Service System (PSS)

I'll just briefly touch on the Passenger Service System. These are usually part of the IFE on Wide-Body aircraft and a completely separate system on Narrow-Body aircraft.

These systems control the reading lights and flight attendant call functions.

- Wide-Body
  - Controls are in the PCU in the passenger seat-arm or on the video touch-screen
  - Signals must be translated and sent to various aircraft devices to annunciate/control the desired action
- Narrow-Body
  - Controls are in the Passenger Service Unit (PSU) above the passengers head
  - Always supplied by the OEM

Figure 22 – Narrow-Body PSS

Business Jet PSS systems are tightly integrated into the Cabin Management System (CMS) along with the IFE and will be detailed later in this book.

Office-In-The-Sky (Internet, E-mail, Etc.)

Here is where we start getting away from "traditional" IFE systems and feature sets. Although now some could argue that this functionality is a "must have" on nearly any flight of any length.

- So far the biggest push has been over-water wide-body aircraft. (Long-Haul flights)
- However, inroads have been made into narrow-body and special route segments (Short-Haul Flights)
- Airtime cost is probably the biggest hurdle

The cost of the air to ground transmission (via satellite) has been the biggest hurdle in the past, and still remains a key obstacle to wide-spread adoption.

However, within the US airspace the implementation of direct air-to-ground communications has been implemented which cuts the airtime costs to a fraction of what satellite time costs.

Couple that with the high-speed broadband connection, and the overall costs in the direct air-to-ground communications systems become almost insignificant.

One other avenue airlines have tried in order to reduce the airtime costs, is to try an approach whereby prior to the flight they store a bunch of content downloaded from the internet into on-board servers. Then passengers access that on-board data instead of actually communicating with the real internet.

Another issue with the off-aircraft communications is bandwidth, and how many "simultaneous" users the connection can support.

Earlier I talked about how much bandwidth different functions consume, so if you refer to that chart (Figure 10) you should be able to recognize the difficulty in providing 100 passengers with a reasonable experience if they are all trying to download and watch HD YouTube™ videos. Especially if the pipe to the aircraft is only an older 640Kbps satellite connection.

And then there are the security issues. How do you protect people from others on the aircraft? How do you keep hackers out of your laptop? Keep you firewall up to date, and hope that the airline has implemented good security measures in their system.

Additionally, in a wireless network, how do you protect the aircraft and data while the plane is on the ground? What prevents someone in the terminal next to the aircraft from accessing the on-board servers? The satellite antennas' ability to transmit on the ground is usually disabled, so that's not really an issue.

As far as implementation goes, like most things IFE related, there are several ways to do it:

- Dedicated "Stand Alone" System
    - Not part of traditional IFE infrastructure
    - Dedicated boxes and cabling (and/or antennas)

- Wired or Wireless
- Integrated into the IFE System
  - Every major IFE vendor offers this capability
    - PROVIDED there is a communications link off of the aircraft
    - Not all of them have actually FIELDED this capability
  - Both wired and wireless options are available

Now, I mention that there are both wired and wireless versions, in most cases only the wireless version is implemented. However, wired options are usually available without changing the IFE infrastructure.

Stand-Alone Wireless Office-In-The-Sky

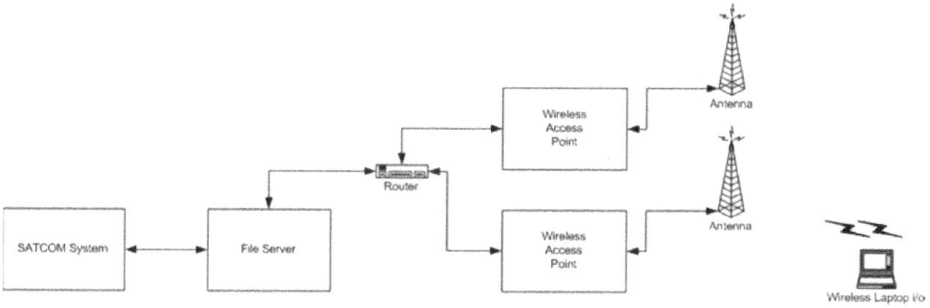

Figure 23 - Stand-Alone Wireless Office-In-The-Sky

With reference to Figure 23:

- SATCOM here represents almost anything to get data on and off the aircraft. It could also be the direct air to ground broadband network or cellular 3G/4G networks, etc.
- WIRED access could also be done at the router or it could connect to other router/switches or the IFE seat boxes or etc.

The big thing to watch for here with wireless is the need to overlap nodes and compensation for dead spots.

Also, while not recognized in the US as a problem, many European countries are starting to investigate the purported health risks/issues associated with long and short term exposure to these Radio Frequency (RF) transmissions.

So, with an "always-on" high-speed broadband connection you can have:

- Internet Radio
- Internet TV
- News, sports & weather
- Unlimited entertainment potential
- And of course productivity tools
    - E-mail
    - Research
    - File Transfers

However, with this always-on connection do you have to think about "undesirable" content? What about access to "adult" web sites while in-flight? Will the airline act as a Nanny and regulate what the passengers have access to?

What about the question of who is acting as the Internet Service Provider (ISP). Is it the airline or the off-aircraft communications company?

Lufthansa "Flynet"

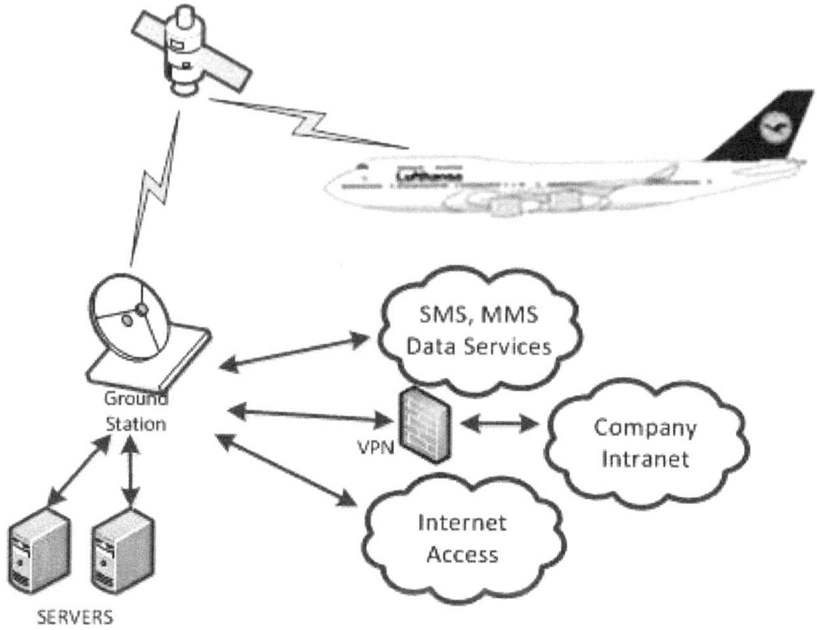

Figure 24 - Lufthansa "FlyNet"

One airline that offers this service as a stand-alone product is Lufthansa with its "FlyNet" product.

The costs seem pretty reasonable for what you get at 30,000 feet above sea level.

- A one-hour flat rate of: 10.95 euros (or 3,500 frequent flyer miles)
- Or a 24-hour flat rate of: 19.95 euros (or 7,000 frequent flyer miles)
- Broadband (scalable):
  - 6 Mbps (into aircraft)
  - 1 Mbps (off-aircraft)
  - Keep in mind this bandwidth is "shared" between ALL the users, including the cabin and flight crew

- Being implemented onto the entire Lufthansa long-haul fleet
- Transmission mode: satellite-based broadband

There have been several systems like this flying for a while and Panasonic released some of the usage statistics from May and June of 2011.

As you can see in Figure 25, the average utilization per passenger/user was 95MB. With the trend definitely moving toward higher and higher data utilization.

With the peak of 155MB utilization per passenger happening at the end of the released data. We can only assume that this trend continues as passengers and crew utilize the services more and more and awareness grows.

In-Flight Internet Data Usage

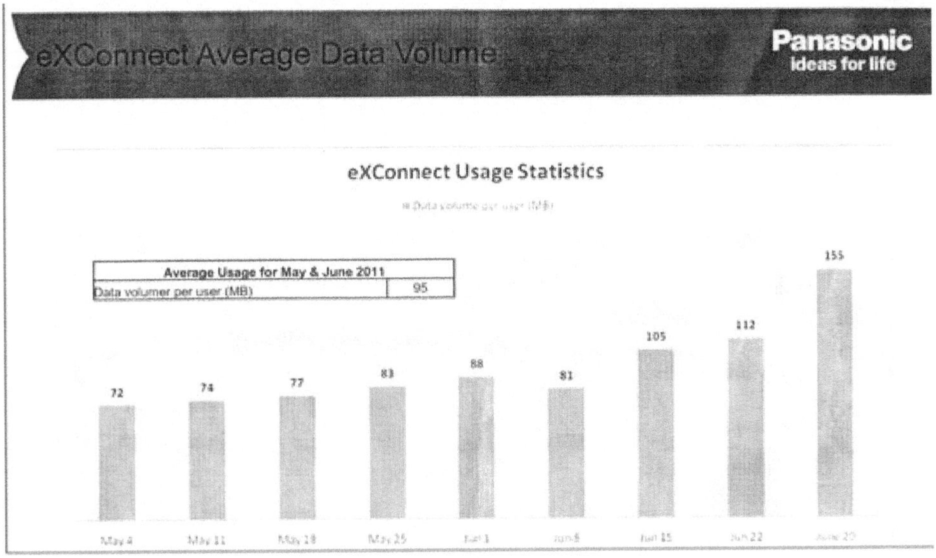

Figure 25 - Panasonic eXConnect Usage Statistics

As far as updated bandwidth numbers go, the last I heard was that Panasonic was claiming 32Mbps data rate available from the satellite to the aircraft. This bandwidth is of course to be shared among all the passengers, the crew and aircraft data needs. What is not clear however is if this is dedicated, and guaranteed, per aircraft, or if this is the maximum that is available in a specific geographic region and is then shared among all aircraft flying in that region.

Viasat SATCOM

Viasat SATCOM is another suppler with a proven solution. Used in many military aircraft, they can provide a 10 to 30Mbps downlink and typically a 4 to 10Mbps uplink. No concern about someone listening into your data communications either as they protect their IP traffic with HAIPE® Type 1 encryption.

Row 44

Row 44 is another major player in this market with a large number of Ku-Band installations on many different types of aircraft. Their IFE offering includes:

- Live Television
- Video-on-Demand
    - Hosted Via On-Board Servers (1.8TB storage)
- Games
- Shopping
- Flight Updates

Ka-Band versus Ku-Band

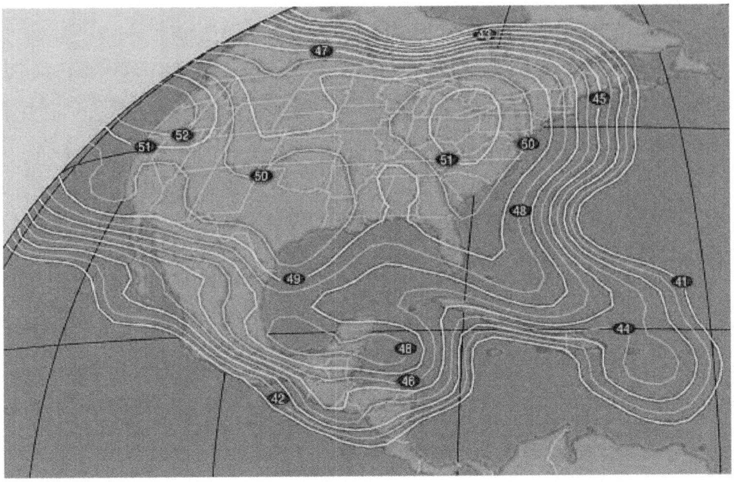

Figure 26 - Ku-Band

Ku-band uses a few wide beams to cover large swaths of land, while Ka-band uses many small beams to cover large tracks of land.

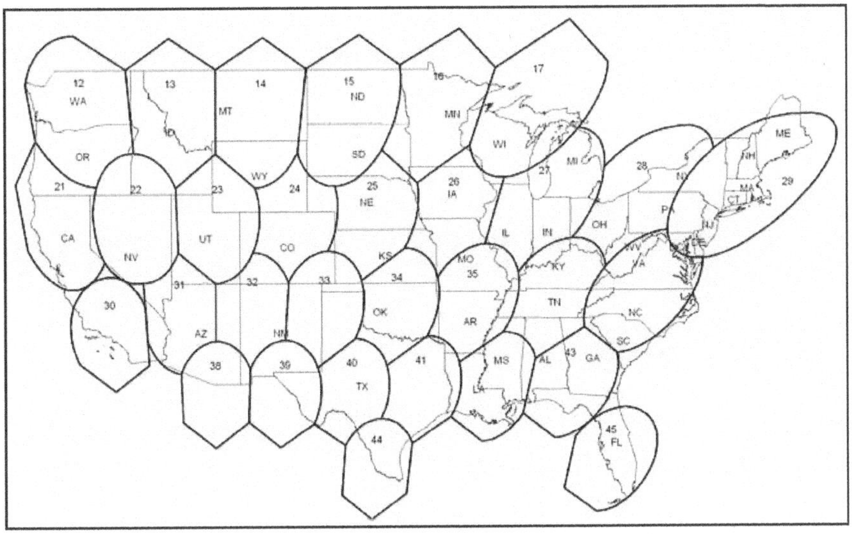

Figure 27 - Ka-Band

Can't really comment on which is "better". They both have good things going for them.

Ka-band generally has much higher theoretical data rates and bandwidth than Ku-band systems. This of course is what everyone is striving for.

But in an aircraft there will be a lot more disconnects as you travel from beam to beam.

Ka-band beams are not very good about handoffs to the next beam. And for that matter, neither are Ku-band, but the problem is exaggerated by the sheer number of Ka-band beams.

Now, while you are driving in your car or walking along a trail in the mountains that isn't really an issue as there is a lot of time for the hand-off to occur.

However, at 30,000 feet and flying at 600+ miles per hour, the hand-offs are typically not very pretty and there are lots of opportunities for dropped connections. Couple that with a plane flying through maybe 13 hand-offs on a cross country (USA) flight and there are a lot of opportunities for passenger dissatisfaction with a Ka based system.

So the trade-off right now seems to be, faster connection speed but lots of disconnects versus a slower speed with few, if any, disconnects.

Terrestrial Broadband

I mentioned Terrestrial Broadband earlier but thought I'd say a few more words about it.

The Federal Communications Commission (FCC) in the USA auctioned off frequency spectrums for high-speeded air-to-ground data networks. The winner of this auction was AirCell (now GoGo) for the primary spectrum license.

GoGo has since fielded hundreds of ground stations pointing upward to communicate with aircraft fitted with their systems.

- Estimated GoGo data rate is 3Mbps on and off the aircraft
    - Used for data and accessing email, web browsing and other related activities
    - Used for Voice over Internet Protocol (VoIP)

It's interesting to note that the secondary spectrum license went to LiveTV. However, LiveTV has yet to utilize any of the spectrum that they purchased.

Additionally, GoGo has recently added a service (GoGo 2Ku) that sends the request for data from the aircraft through the Air-To-Ground (ATG) link but the data response to the aircraft is from a satellite connection at a much higher data rate. This keeps the antennas very small and the connections very fast.

In-Flight use of Cell Phones

I'd like to start off by saying I really hate this idea. It's bad enough when I have to sit next to someone in a restaurant and they are yelling into their cell phone, I don't want to have to sit next to someone for 12 hours while they shout non-stop into their phone and I have no opportunity to escape.

Have you heard the term "Road Rage"? Now couple that with a half-dozen beers or shots of whisky and let's see what happens when passengers want to sleep and their seat mate won't shut up. AIR RAGE!!!

Technically speaking, putting a pico cell into an aircraft and getting a call to the ground is pretty darn simple. The first test installation was over 10 years ago.

As you can see in Figure 28 there are typically only a handful of components. Some vendors add a few more but essentially it's only:

- Base Transceiver Station (BTS)
    - Ethernet Switch and Server
- Radio Network Control Unit (RNCU)
    - GSM Radios
- Antennas

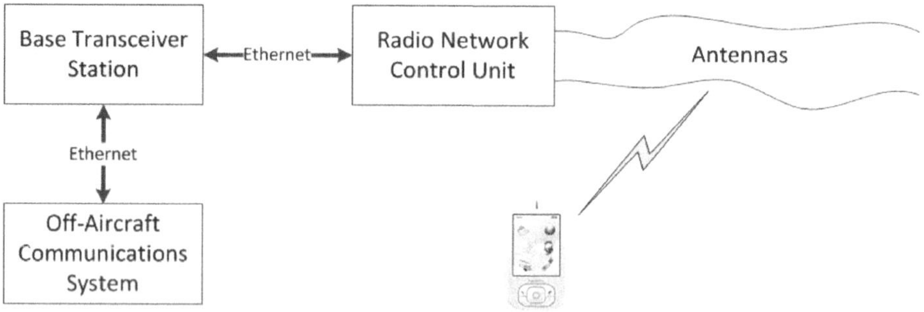

Figure 28 In-Flight Cell Phones

The way this works is that your cell phone connects to the "Cell Tower" in the aircraft and since it senses a cell tower so close, it reduces its power output to its lowest level. That's what makes it okay with many government agencies as it's not going to blast a hundred cell towers on the ground. The signal will be too weak to reach them.

Regulations about Cell Phones in Aircraft

In the USA, most people think that it's the Federal Aviation Administration (FAA) that specifically regulates cell phone use in aircraft while airborne, but that is not true, it's the Federal Communications Commission (FCC). Specifically, the "Code of Federal Regulation (CFR) Title 47, Part 22: Public Mobile Services".

- ✈ Sec. 22.925 Prohibition on airborne operation of cellular telephones.
  - Cellular telephones installed in or carried aboard airplanes, balloons or any other type of aircraft must not be operated while such aircraft are airborne (not touching the ground).
  - When any aircraft leaves the ground, all cellular telephones on board that aircraft must be turned off.
  - The following notice must be posted on or near each cellular telephone installed in any aircraft: *"The use of cellular telephones while this aircraft is airborne is prohibited by FCC rules, and the violation of this rule could result in suspension of service and/or a fine. The use of cellular telephones while this aircraft is on the ground is subject to FAA regulations."*

There are several countries, most notably in Europe, that have lifted their ban on the use of cell phones in aircraft; the USA has steadfastly kept the ban in place.

Cell Phone Connectivity

Some of the companies operating in this cell phone space are listed below:

- AeroMobile
    - Operates via classic Inmarsat Aero-H/H+ satellite communications systems already on board 1,900-plus long-haul airliners.
    - Estimated roaming charge is $3.50 per minute
- OnAir
    - Works with GSM phones only
    - Operates via Inmarsat Swift Broadband
        - Also allows for email and internet connectivity
        - Estimated price of $9.95 per flight
- GoGo (Aircell)
    - Allows both GSM and CDMA phones to be used
    - Target cost to airlines for the equipment is less than $100,000USD per aircraft.
    - Internet access estimated to cost $10 per flight
    - Estimated roaming charges for voice services are approximately $0.25 - $0.30 per minute.

Cell Phones in Aircraft Fleets

Airlines are installing these things right and left as fast as they can. As I'm writing this, I believe the following is where the industry stands with regard to installations:

- Emirates: B777 Fleet
- Ryanair: 50+ Aircraft
- Air France: Trial
- Qantas: Trial
- TAP Air Portugal: Trial
- BMI: Trial
- OMAN Air: A330 fleet
- Malaysia Air: B777

- TAM currently installing on
    - 12 B777,
    - 10 A330,
    - 27 A350 and
    - 31 Short-haul (through 2018)

International Airline Passengers Association (IAPA)

The International Airline Passengers Association (IAPA) has published a survey revealing that there is widespread opposition from passengers to impending plans to introduce mobile phones on aircraft. Those polled fear onboard telephony would be disruptive, anti-social and possibly even dangerous.

...the survey found that although 90% of IAPA members who took part in the research carry mobile phones when traveling, only half of the respondents said they would find it useful to be able to make calls during a flight, and many were violently opposed to the development. *From Flight International Magazine, 28JUN06.*

The airlines of course are clamoring for the approval of cell phones in-flight in the US. Why do you think that is?

Easy, REVENUE! The airlines get paid part of the roaming fee your cell phone carrier collects from you for using your phone out of your "home" area. The airlines expect to make hundreds of billions (yes Billions) of dollars from these roaming charges.

So they don't really care if passengers get mad at each other, as long as they are addicted to using their cell phones. The airlines get paid for voice calls, texting and internet roaming so it's easy money for them.

In-Seat Power Systems (ISPS) (Laptop Power)

When Laptop Power Systems first began being fielded, they were +15Vdc systems. You used an adapter like for your car cigarette lighter, which was something most road warriors had at the time.

The original in-seat power company went through a host of company name changes in the beginning (from Olin-Aerospace, to PRIMEX, to General Dynamics, to Astronics) as this company was purchased by that company which sold off that division which was then sold... You get the idea...

Also near the beginning was Page Aerospace which was the first in-seat power competitor to PRIMEX. Page had a variable voltage system that relied on a "smart" cable between the laptop and power supply to tell the power supply what voltage it should output. Unfortunately PRIMEX had already captured a huge percentage of the market and eventually at an airline industry meeting Page endorsed the fixed voltage +15Vdc as the Airline standard for In-Seat Power. Unfortunately, Page was never able to gain enough market share to sustain in-seat power as a product line.

There are currently two major players in the in-seat power market, Astronics (with over 235,000 seats installed) and KID-Systeme, a division of Airbus Industries.

Also, currently there are two flavors of in-seat power, the original +15Vdc and now 110Vac 60Hz. Kid-Systeme was the first to field the 110Vac system but they were not the first to develop it.

Way back in the beginning, Olin's initial in-seat power design was for an 110Vac 60Hz in-seat power system. However, a certain executive within the FAA is reported to have said, "Over my dead body will an AC system be allowed for in-seat power". So, Olin had to change their design to DC and the rest is history.

Apparently, by the time KID-Systeme brought their design to the FAA, that individual had moved on to other things, as they were given a list of criteria to meet and they met it and got their system certified. Very shortly thereafter, PRIMEX dusted off its initial designs, made a few changes to meet the new FAA criteria and they too had an 110Vac system for sale.

Currently, all new in-seat power systems that are being installed are AC.

In many cases these systems are installed as stand-alone retrofit installations. This is a very inexpensive way for an airline to breathe new life into their IFE offering without spending an arm and a leg.

Most new aircraft installations are done as an integrated solution as part of the IFE. Either incorporating the ISPS functionality into one of their LRUs, or, adding it as a parallel power system that also provides power to their in-seat IFE boxes.

Some of you may be thinking, "AC Power? But all the plugs around the world are different." And for those of you thinking that, congratulations, that was an interesting obstacle. So what they developed were single outlets that could take the plugs from 140+ countries. Figure 29 shows a typical multi-country outlet unit; note the unique shapes of the two primary sockets.

Figure 29 – Astronics Multi-Country In-Seat Power Outlet

The certification requirements between the DC and AC systems are also vastly different. For example, the AC system requires an "In-Use" light on the seat and whenever someone plugs something into the outlet, the in-use light comes on. The airlines are then required by the FAA to have the Flight Attendant go to the seat and look to see what the passenger has plugged in and to ensure that it is an authorized device. So no soldering irons, hair curlers, TV sets, radios, etc.

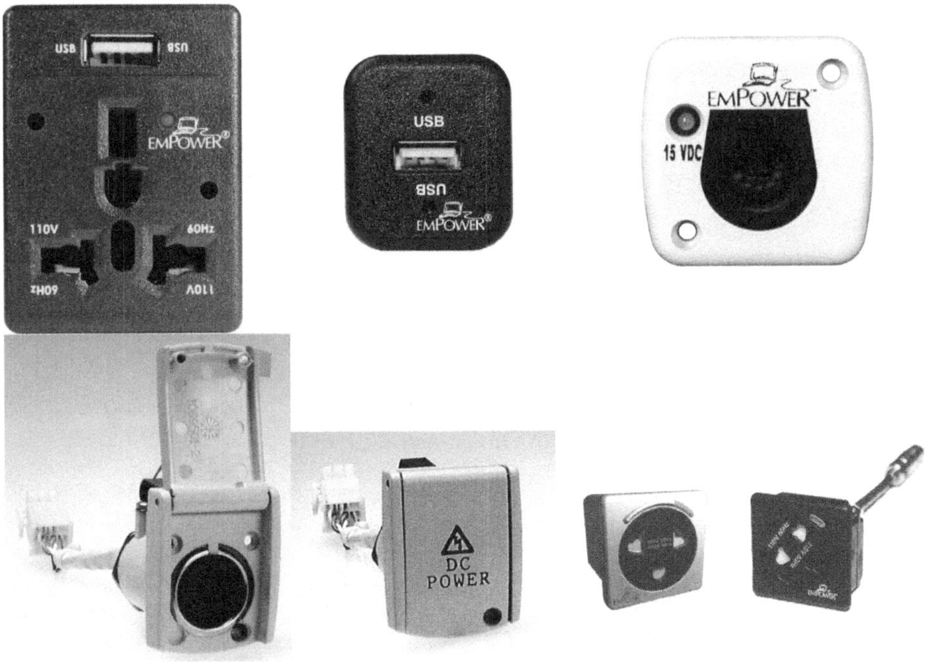

Figure 30 – Some Additional Astronics In-Seat Power Outlet Types

Typically the outlets themselves have no power applied until a proper plug is inserted. Once a proper plug is inserted the Outlet senses that and then the power supply applies power to the Outlet. This is to prevent arcing and sparking in the Outlet as plugs are inserted and removed.

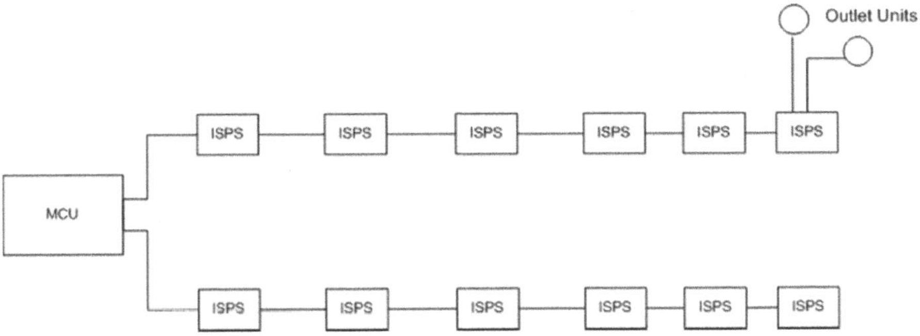

Figure 31 - Stand-Alone In-Seat Power

When in-seat power first came out, 100 watts was more than any laptop out there needed and that's what the FAA said they would allow. Fast forward to today, and laptops are power hungry beasts. So, the FAA recently upped their allowed power outputs to 200 watts per passenger, which, for the time being should satisfy all but the hungriest of laptops.

Astronics

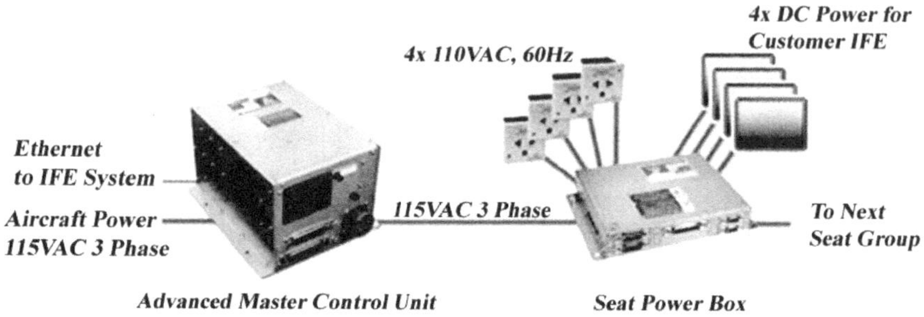

Figure 32 – Astronics In-Seat Power Plus IFE Power

Kid-Systeme

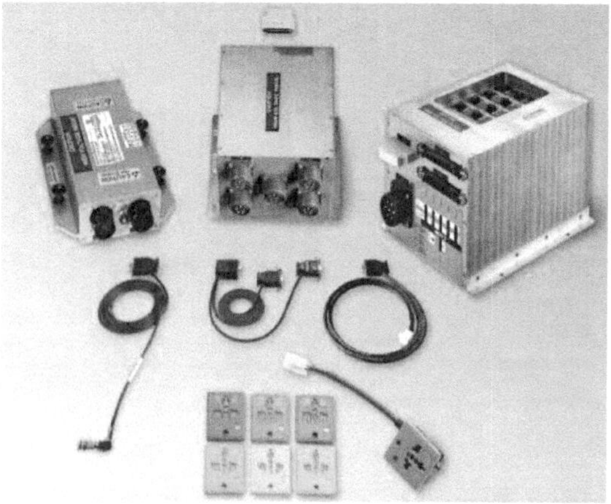

Figure 33 - Kid-Systeme Hardware

All of these systems regulate the amount of power that the passenger equipment can consume from the aircraft via active power management methods. Currently, Astronics holds many US Patents in the "active power management" arena.

For example, if an aircraft has 300 seats fitted with in-seat power, and they are all consuming 150 watts, that would be 45,000 watts of power that they would be trying to draw from the aircraft.

Unfortunately, no aircraft flying today has that much spare capacity to give up to the in-seat power systems. So, the systems are set with a maximum power budget they can take from the aircraft and once that budget is reached, all outlets not currently in use are switched off until such time as the power load diminishes.

And of course with the increase of USB powered devices by the flying public, there has been a move to only providing USB power ports in many aircraft, as opposed to installing the AC outlets. Although on quite a few installations now the operators have opted to include both USB and AC power.

Video Games

Who doesn't love video games? Okay, I admit there are a few people out there, but with the video game industry sales dwarfing that of the traditional movie industry, I think the majority of people enjoy them.

Aircraft systems typically provide a wide variety of games depending of course on the system installed, the route, geographic location, airline desires, etc.

This section will deal with the general categories of games and not necessarily who is installing them.

We typically split the games category into three (3) general categories:

- Simple (Hangman, Solitaire, etc.)
- Complex (Nintendo, Playstation, etc.)
- Multi-player (LAN games)

If available, these are typically offered as part of the IFE system. A complex tethered PCU is used typically to control the game functions. However, with many airlines installing purely seat-back system with touch-screens, the PCU has been eliminated for some of these carriers.

Briefly, let me mention "Portable" gaming devices. Portable gaming devices HAVE been used in aircraft before, (from the history section earlier: Atari in 1975), but problems with batteries, theft and breakage meant they didn't last long as a product offering in the aircraft.

Simple Video Games

Simple video games are, well, simple. They can be easily started and stopped, and there is typically little to no emotional investment. Simple games include games like:

- Solitaire,
- Hangman,
- Reversi,
- Memory challenges,
- Mahjong,
- Tetris,
- Bejeweled,
- Sudoku and Etc.

Complex Video Games

These games take a bit more effort and are sometimes played against other passengers and not just the host computer. These games types include:

- Super Mario Brothers
- Nascar
- Donkey Kong and Etc.

At the time of this writing, Qantas has over 80 games available in its IFE offering. A huge mix of simple to complex gaming options.

One of the issues here though is that unlike the simple video games there is some emotional investment typically associated with the more complex games. And many passengers will also be playing these at home as well as in the air. But as time marches on, these games can become a bit stale and depending on the system, it's not always that easy to update them.

There is also the issue of the handheld keypad control. The Passenger Control Unit (PCU) needs to be pretty complex for these games to be playable. Now, I know some of you are thinking just use the touch-screen interface on the monitor. But remember you'll be poking the back of the seat of the passenger in front of you and they may not take too kindly to that aggravation.

And then there is screen size. Most of these games were designed to be played in your living room on a decent sized television. When translating down to the small aircraft monitors, they are not always as pleasurable to play.

Licensing is another area of concern. These hot popular "branded" games are expensive to license. And their costs are passed on to the passengers by way of higher prices.

I mentioned game obsolescence as it's a real issue. How many games are you still playing that you were playing five years ago? I'd wager not very many...

So ease of updating the games has to be a high priority. And many of the newer IFE systems use software emulation modes which are much more tolerant of game changes and updates than using the more traditional embedded hardware solutions.

Which is a nice segue into the world of on-board gaming: Multiplayer, Local Area Network (LAN) Video Games.

Multiplayer (LAN) Video Games

This to me is the Holy Grail of gaming in an aircraft.

Yes, there are certainly multi-player options in the "complex" games listed earlier, but nothing beats the complexity and immersion of industrial strength multi-player LAN games.

This category encompasses several genres of multi-player games such as:

- First Person Shooters
- Strategy
- Role Playing
- Sports
- Action

Gaming franchises such as:

- Quake
- Half-Life
- Doom
- Call of Duty
- Age of Empires
- FIFA

Hosting these games on an on-board server is one option, and so is playing while connected to a high-speed off-aircraft broadband link to servers on the ground. I'll talk about that second option later.

Another option is to allow players to host the campaign on their own computer and allow other passengers (players) to join in via the network. This scenario is typical of many LAN parties at homes across the world.

However, one thing to keep in mind would be security issues. If a player is connected to the network, is her laptop open for an onboard hacker? In most cases anyone with a laptop, tablet or other Wi-Fi enabled device has taken the appropriate steps to protect it. Certainly in the gaming community players are much more cognizant of the risks of on-line activities then perhaps the average e-mail or Facebook user may be.

So far this hasn't taken off very well yet. Perhaps because of the licensing costs, or the perceived limited demographic. Whatever the reason, I'm still waiting for this to get traction and have some widespread adoption.

I know many people that can sit down to play one of these video games and play non-stop for 8 to 10 hours. This matches the flight time of a long-haul flight pretty closely. A perfect way for them to pass the time and not get anxious about flying.

Headphones

Today's headphones/headsets come in a variety of types, sizes and quality. I mentioned the old air tube or pneumatic style of headset before but to my knowledge nobody uses them anymore and hasn't used them for years. So, I'll focus on the electrical headset variants.

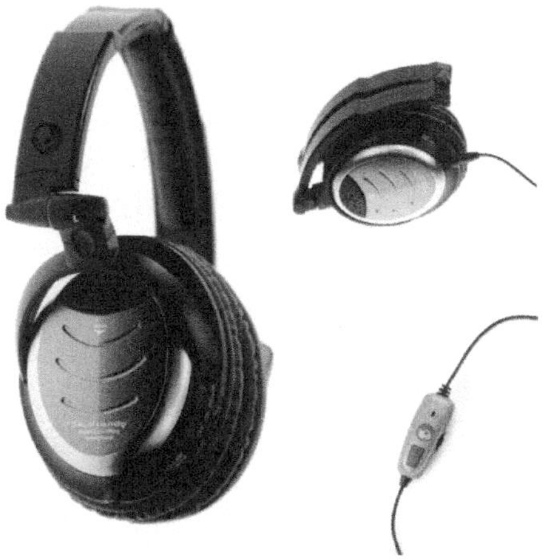

Figure 34 - Headphones

When the IFE systems moved away from pneumatic to electrical headsets, the immediate questions that arose were what plug styles to use and what impedance to use.

Plug styles are quite diverse with single stereo plug, multiple mono-plug, stereo + power plug and tri-plug designs common. The impedances vary from system to system and can be anywhere between 32Ω and 300Ω.

So, in the professional headset market, the headsets tended to use higher impedance headphones with higher power inputs. This meant in the early days of electrical headsets and IFE, the IFE vendors tended to use 300Ω impedance headsets. As a point of reference, typical consumer headsets are only in the 32Ω to 40Ω range.

Also, since the airlines charged passengers, in coach, for the use of the headsets, they needed a way to discourage people from using their own.

That's why many airlines opted for the two-prong headset jack with left channel on one plug and right channel on the other. Since virtually all consumer headsets used a single 1/8" stereo plug, this made it difficult for passengers to use their own headset.

What made it more interesting is that in the USA we used a plug spacing between the two mono plugs of 0.5" (12.7mm) whereas in Japan and other parts of Asia they wanted a plug spacing of 0.531" (13.49mm). This meant that in addition to the IFE vendors having to make unique PCUs/headset jacks, that the aircraft headset manufacturers also had to design different headset plug styles to match those being installed on different airlines around the world.

Not confusing at all right?

Headphone Plug Styles

This is by no means a comprehensive list but it gives you an idea of some of the different shapes and styles.

- Standard
  - 1/8" single stereo
  - 1/8" + 1/8" dual mono prong
- Noise canceling
  - Dual prong (1/8" stereo + 3/32" power)
  - Tri prong (dual mono prong + 3/32" power)
- Right angle or straight plug?
- Headset Jack integral to the PCU or remote?
  - One part number or two?

- Keep in mind that the headset jack fails more often than the PCU so there are advantages to having it be a separate piece.

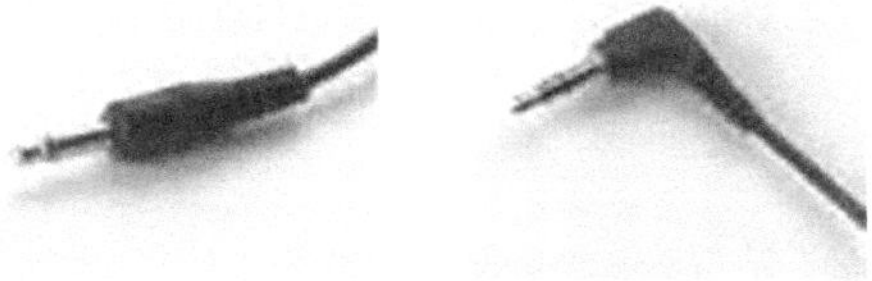

Figure 35 - Headphone Plug Styles

Today, most airlines use the single stereo plug style headset jack. Although there are still some unique ones flying but those usually, but not always, have to do with noise-canceling headsets which I'll get to in a minute.

Figure 36 - Headphone Plug Styles

Also today, we see more and more airlines just handing out inexpensive ear-bud type headsets to passengers and telling the passengers to just keep them. In today's world, this is usually more cost effective for the airlines than having the Flight Attendants collect the headsets, sending them off to be cleaned, tested, re-bagged, stored and then sent back out to another aircraft.

Noise-Canceling Headsets

There are several types of noise-canceling headsets that have been used in aircraft as part of the IFE system.

The three key areas that differentiate the different designs are:

- ✈ Source of Power
- ✈ Circuitry Location
- ✈ Microphone Location

Power

For those IFE systems that are designed to have a noise-canceling headset power output, the IFE industry has defined this voltage as 13.5Vdc. This is provided via a special Noise-Canceling headset jack. This jack provides power in addition to the left and right audio channels.

And of course there are headsets in use that are powered by AAA or AA batteries. But these are few and far between as the costs of maintaining these are fairly high.

Circuitry Location

The circuits that actually do the noise cancelling can be located in several places, each with their own advantages and disadvantages.

- ✈ In the Headset

This placement lets the headset vendor build a completely self-contained unit, like one you would buy yourself. Unfortunately, having the electronics in the headset tends to make it a lot more expensive option.

- ✈ In the Headset jack

This placement lets the headset vendor build an inexpensive headset as it moves the cost to the jack. On the down side, it makes the headset jack more expensive and a little larger.

However, one company has an innovative solution: InFlight Peripherals Limited has a unit they call INCAM™, for Integrated Noise Cancelling Audio Module which is a pretty neat device. This can significantly reduce the costs of having a noise-cancelling system installed and their rapid and long life headset jacks are really amazing pieces of engineering.

Figure 37 – Integrated Noise Cancelling Audio Module (INCAM™)

- In the Passenger Control Unit (PCU)

This placement also lets the headset vendor build an inexpensive headset as it moves the cost to the PCU. It tends to have a minor impact on the PCU price but typically no impact to the PCU size.

Most vendors and airlines have opted for the electronics to stay within the headset.

Microphone Location

The microphone placement is essential to the effectiveness of the Noise-Cancelling headset being able to actually negate the ambient aircraft noise.

The options are basically the same as for where the noise-canceling electronics are located.

But first a little primer on how Noise-Canceling works.

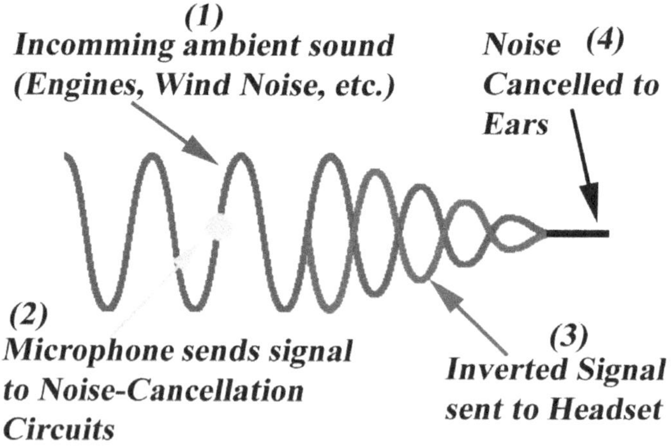

Figure 38 – How Noise-Canceling Works

Noise canceling headsets work on the principle of canceling out a constant ambient noise. Like the hum of a fan, or engine noise, or wind noise in an aircraft.

Please reference Figure 38: the ambient, repetitive, sound **(1)** is picked up by a microphone **(2)**, and the microphone sends the signal to the electronics which inverts the signal and injects it into the audio stream **(3)** heading toward your ear **(4)**. The effect is that this new signal cancels out the original signal and the repetitive noise is eliminated.

Recently I've seen some ads and commercials for noise cancelling headsets that show an aircraft passenger cabin with people talking loudly near a passenger and then the passenger puts on a noise-cancelling headset and then blissfully listens to his music.

Unfortunately, the reality is that this is exactly the opposite of what would actually happen. You would hear these people better, not cancel them out, as you would be cancelling the aircraft's engine and wind noise, NOT the people talking.

So, on to the possible microphone locations:

- In the Headset

This is the best place for the microphone as it's located next to the ear. This is important as the noise-canceling circuits' best cancel noise when the input signal is closest to where the noise is entering the ear. Anyplace else and the sound wave will be slightly out of phase and the canceling will not be as effective.

✈ In the Headset jack

Some people put the mic inside the headset jack when they place the electronics there. Not an optimal location but it can be made to sort-of work.

✈ In the Passenger Control Unit (PCU)

Another non-optimal place to put the mic. The hard part here is that depending on the PCU location, inside the seat arm side for example, the ambient sound may be too attenuated to effectively sample and re-inject into the audio stream to get a good noise-canceling affect. And again, it's located too far from the ear to have an optimal noise-canceling operation.

iPod Jacks

Figure 39 - iPod Jack

These are special connectors that have a small DIN type jack. The passenger would need to use a special adapter cable, obtained from the Flight Attendant, to connect their iPod to the IFE system.

This lets the passenger view iPod content on the larger in-seat video monitor and also to listen to their iPod music through the headset connected to the IFE system.

Now, while that may seem a bit silly, why connect the iPod to the IFE to listen to audio, it's handy when you want to switch between the on-board music selection and your own. But more typically, it's there so that you can watch your iPod video on a larger screen.

Passenger Control Units (PCU's)

These have gone by many names over the years. From Passenger Control Units (PCU), to Digital Passenger Control Units (DPCU) and Advanced Digital Passenger Control Units (ADPCU) and other such variants. For simplicity, I'll just be referring to them as PCUs.

These are the units that are typically located in the passenger armrests and is where the Human Machine Interface (HMI) is located. Even wide-body IFE systems that have in-seat video monitors still typically also include a PCU in the seat.

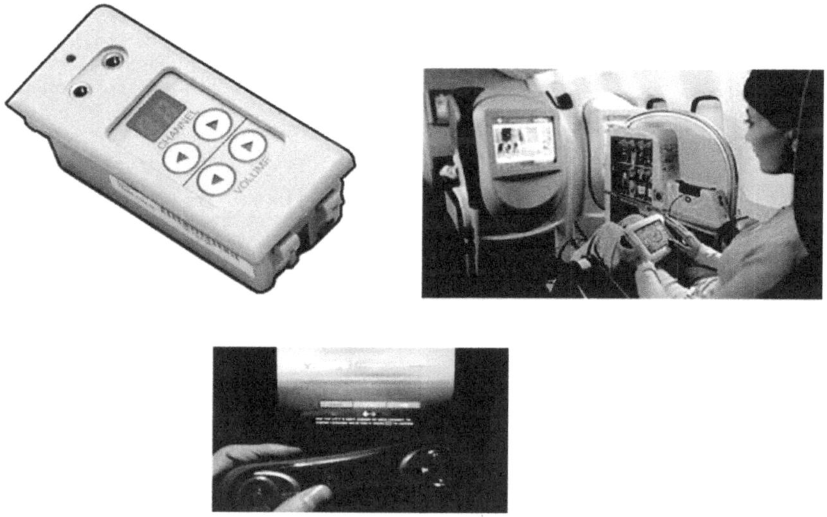

Figure 40 - PCU's

The amount of functionality these can have in them is staggering. From simple audio channel selection and volume controls to complex devices with keys and buttons and swash-plates on every available surface.

The following pictures show a PCU with 68 buttons, it's capable of controlling a huge array of IFE and Cabin Management functions.

Figure 41 - Handset Front

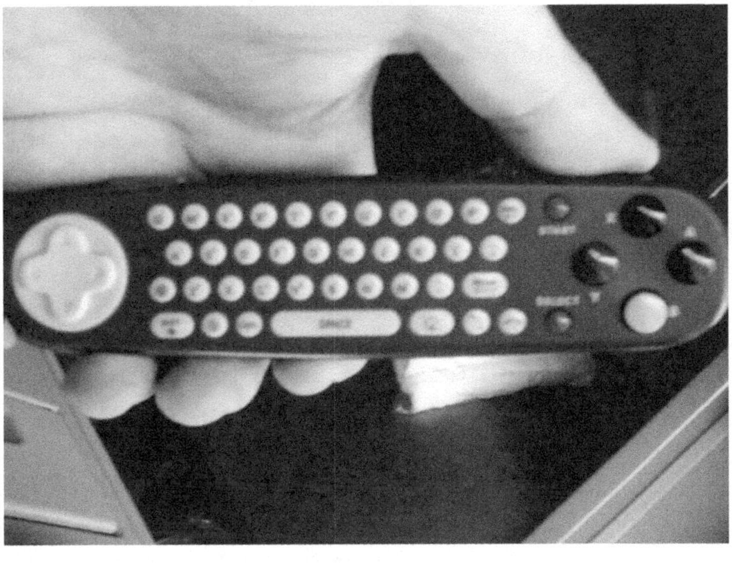

Figure 42 - Handset Back

Currently the trend is to move away from these kinds of controllers with dedicated hard buttons and move toward a more universal touch-screen device like a "smart" cell phone.

In that way you no longer need to pre-define all the button functionality and have different hardware part numbers for every client. You could standardize on only a few part numbers of units with different sizes or finishes. All the functionality could then be realized, and rapidly changed, from client to client via software.

Figure 43 – Smart IFE Controller

Additionally, the complexities of the PCU and its interface have to be considered. Don't make the units a challenge for passengers to use.

I was once on a flight where the PCU was tethered, it was about 7 inches (178mm) long and had buttons covering three sides. One side had numerous gaming buttons, swash plates, the Reading Light, FA Call, audio and channel select buttons and the other main flat side had telephone keypad and a full QWERTY computer keyboard. One of the long thin sides had a few more gaming buttons while the other thin side had a credit card reader.

All totaled I think this unit had over 140 discrete buttons on it. Nearly double the amount of buttons shown a few pictures back.

That particular flight is memorable to me because of the following experience. On that flight I recall a rather elderly woman sitting across the aisle from me and she was in tears. All she wanted to do was turn on her Reading Light so she could read her book, but the PCU was too confusing for her to operate. So I showed her which button to press for the light and how to work the music controls and she was able to finish the flight with a better experience.

This was a PCU with lots of physical buttons, but the same concern can be voiced for touch-screens and their Graphical User Interface (GUI). If not done properly, it will alienate the passengers.

System Installation Considerations

Okay. So now you know "what" you want to install, so where does it all go in the aircraft?

There are basically three (3) general areas we talk about when referring to IFE systems:

- In-Seat
- In-Cabin
- Head-End

With the general definitions being:

- In-Seat = Physically located on or in the passenger seat
- In-Cabin = In the passenger cabin but not on the seat. Generally behind or on sidewall panels; on or above the overhead/ceiling panels; on partitions walls; in, on or under bag bins/stow bins; etc.

- Head-End = Almost anywhere but typically WITHIN the pressure vessel. This is most commonly in the aircraft EE-Bay, in the belly of the aircraft, but could also be in or above the passenger cabin.

Let's start with the In-Seat equipment as that's what the passengers play with and where the entire "Passenger Experience" takes place.

In-Seat Equipment

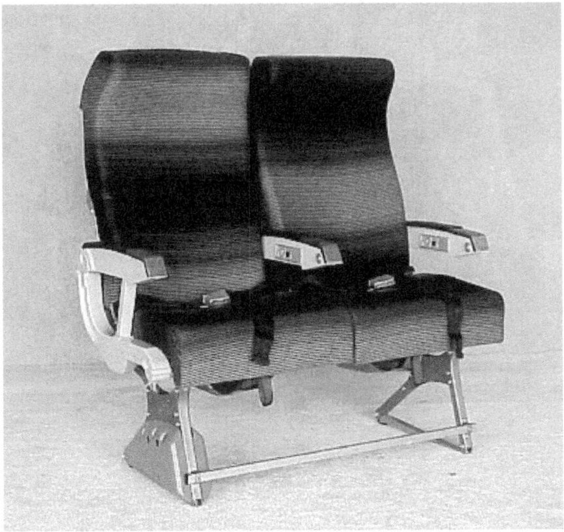

Figure 44 - Double Economy Seat

There are numerous places within the seat that equipment can be placed, which is mostly determined by what exactly that equipment is and what it does.

For example, equipment that the passenger is expected to interact with such as PCU's and In-Seat Video Monitor's, are placed somewhere that the passenger can see and touch them. Whereas supporting equipment, such as power supplies and decoders are buried inside or under the seat structure.

PCU Mounting Locations

Currently in vogue at the moment is the mounting of the Passenger Control Units (PCU) on the top of the seat arm-rest. This location provides an excellent ergonomic position for the passenger to adjust controls with little movement or effort. Typically the PCU is placed as far forward on the arm-rest as possible and where the passengers hand would normally rest in a relaxed sitting position.

Figure 45 – Top-Mounted PCU

Prior to the PCU's being installed on the top of the arm-rest, they were mounted on the inside of the arm-rest next to the passenger's leg or thigh. While being a very protected spot for the PCU, it was harder to operate and needed a bit of contortion effort for some people to operate.

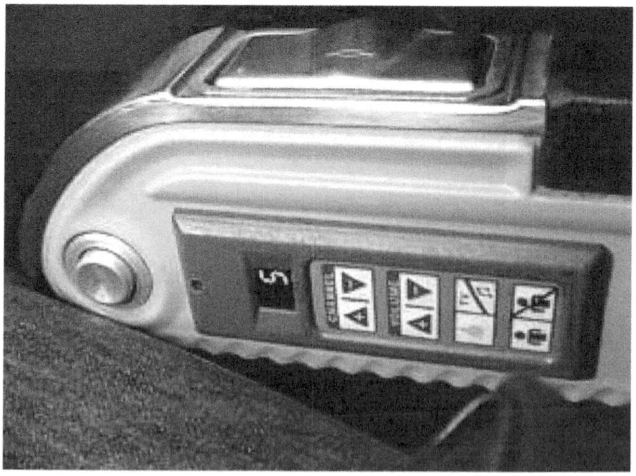

Figure 46 – Side-Mounted PCU

And then of course there are the hidden places for the PCUs to be installed. Sometimes they are placed out of the way under little doors where they are either completely hidden or sometimes the door has a cutout that exposes the most frequently used functions. This is most often the case in First or Business Class seats. It is rarely if ever done in Economy or Coach Class seats.

Also note that these can be either fixed or tethered units. Typically the more functionality being employed the more likely it is that the unit will be tethered so that passengers can pull out the unit and hold it in both hands while operating it.

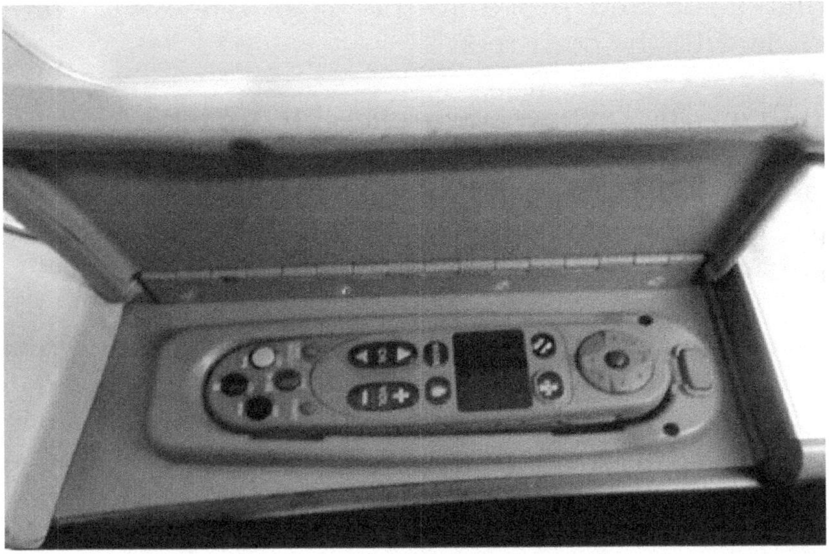

Figure 47 – Under-Cover "Tethered" PCU

PCUs are also occasionally mounted in the seat back. Usually only when there is a video monitor present and then it's installed under, or to the side of, the video monitor.

When one of these tethered PCU's are installed in the seat back, one big issue is with the PCU cable. When someone is playing a game while sitting in the aisle seat, and the window seat passenger needs to use the lavatory, the aisle seat passenger's PCU cable will effectively block their exit.

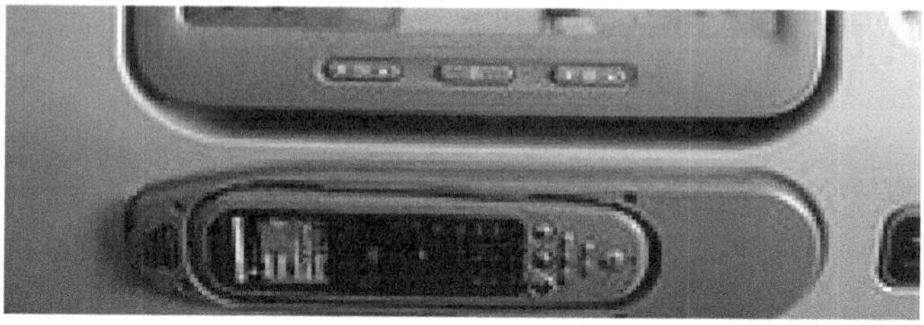

Figure 48 – Seat-Back Mounted "Tethered" PCU

Another issue and probably the biggest one with tethered PCU's, is with the cord-reel.

Or rather, the extreme unreliability of the cord-reel. I understand that in systems that use a tethered PCU, that the associated cord-reel has the absolute highest rate of failure, the lowest MTBF, of any other component in the IFE system.

Now, to be fair, there have been great strides made recently on making these items more reliable. Only time will tell if they have been able to overcome the difficulties in designing and manufacturing these to withstand the collective abuse from millions of passengers playing with them on a daily basis.

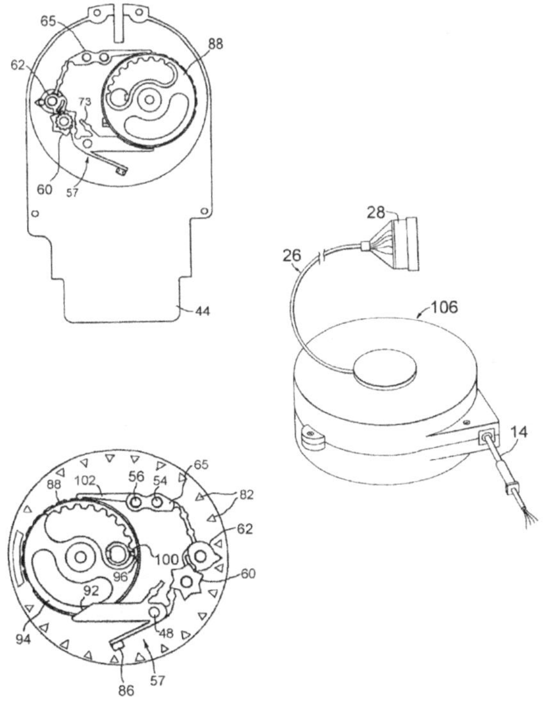

Figure 49 – A "Tethered" PCU Cord-Reel

Typically the more complex the functionality of the PCU the more likely it is to be tethered.

Even many of the newer touch-screen, cell phone like, PCUs can be found on tethers.

Some of the things that system designers and installers need to be wary of when installing the PCUs are: liquid ingress, liquid pooling, drainage, backlighting, day/night illumination and usage.

Liquid ingress, pooling and drainage are big issues with the PCUs. Consider the top mounted PCU, recessed into the arm-rest. When someone spills a drink onto it, where does the liquid go? Is the top completely sealed so it pools covering the buttons? Is the PCU angled with a channel to guide and drain the liquid away and into the passenger's lap? Or does the liquid seep into the unit because the buttons are open? What kind of cleaning solvents can be used on these without melting them?

How are the buttons illuminated for night-time use? Is the illumination dimmable so that it doesn't keep passengers awake during a night flight? What about legibility during bright sunlight? For LCD Touch-Screen PCUs, will bright sunlight washout the screen and make it inoperable?

Side-arm mounted PCUs aren't without their challenges either. In a narrow seat, will the passenger just feel around for the reading light button only to hit the Flight Attendant Call Button by mistake? Ask any Flight Attendant, this happens all too frequently. Also these units have a very limited viewing angle and discussions on which way to face the buttons, forward or upward always make for interesting development topics.

And then there is the cable routed to the PCU and its issues. For seat arm mounted units, the cable is typically routed up from the seat leg, through the movable arm, and into the PCU. As the cable passes through the "elbow" joint of the arm-rest, it is supposed to be connected to a small eyelet so that it stays in place.

Unfortunately, sometimes this attach point gets missed and the cable is left free to move within the seat arm. When this happens, the cable can easily find itself slipping down between the movable portion of the arm and the stationary part of the arm joint. The arm-rest then acts like a scissors slicing the wire. Sometimes cutting the wire cleanly and sometimes causing it to short-circuit to the seat frame. Neither of which is considered a good thing...

In-Seat Monitors

Here is another area where a lot of the innovation has taken place over the years.

Generally speaking, there are three primary areas for the in-seat monitor installations.

- Seat-Back
- Seat-Arm
- Seat-Fixture/Monument

With the general definitions being:

- Seat-Back = The monitor you are viewing is physically located on the back of the seat in front of you.
- Seat-Arm = Your monitor flips out of the monitor stowage area via an arm assembly attached to the seat.
- Seat-Fixture/Monument = Not directly attached to your seat but onto a fixture or monument near your seat.

Seat-Back Monitors

Typically when we think of in-seat video this is the image that comes to mind, rows and rows of seats with video monitors in their seat-backs.

Figure 50 – Seat-Back Video Monitors

Typically these are found in the coach, economy, and tourist class seats. Historically these have tended to be smaller monitors, on the order of 8" or less diagonal. However, as seat designs improve and monitor depths shrink, some seat-back installations have had 10" and larger monitors installed in the seat-backs. Although this is a trend that may not continue as larger sizes impact weight and certification costs. And of course even finding places to physically install them in the newer "slim" designed seats is also a huge issue.

Now, with the monitor installed, the primary issues with this mounting location are the recline angles of the seat versus the typically static angle of the monitor, as well as if the monitors are using touch-screen overlays.

In order to save costs, some seat-back installations use a fixed monitor. That is, as the seat reclines, the monitor moves back exactly in the same position as when the seat was upright.

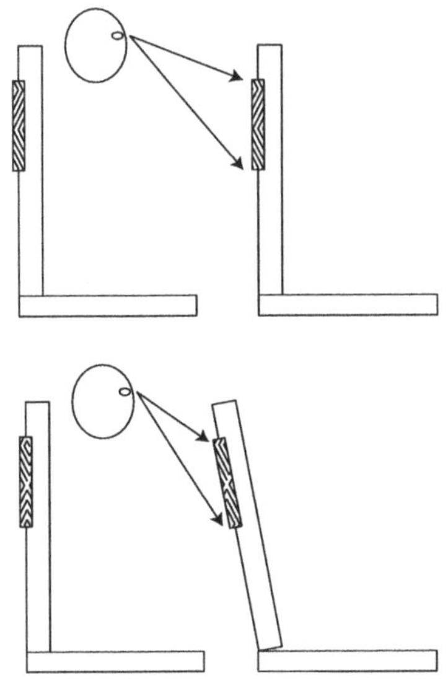

Figure 51 – Seat-Back Monitors: Non-Articulating

Unfortunately, few monitors are able to provide a usable viewing area at such a steep angle. This forces the user to duck down in their seat, usually coupled with them reclining as well, in order to see what is happening on the monitor.

Contrast that with the articulating seat-back monitor design, where the monitor bottom can move out from the seat-back and allows the passenger to position it for the optimum viewing angle

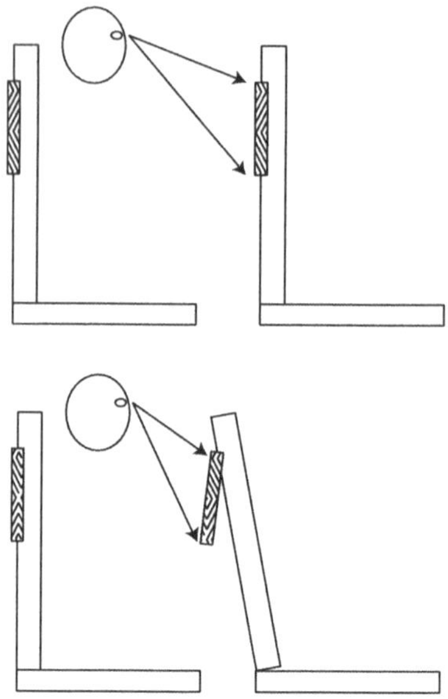

Figure 52 – Seat-Back Monitors: Articulating

The articulating seat-back monitor design allows for a much better passenger experience.

Now, there has been some issue in the past with fingers getting pinched behind the moveable monitor. However, since these are all manually moved by the passengers, there isn't much that can be done here. Some airlines have opted to install small accordion like baffles between the monitor and seat-back to limit the passenger finger intrusion into this area.

Seat-Arm Monitors

These types of installations were first used in First and Business Class seats. As the recline was so much greater than that of an economy class seat, it would be impossible for the passenger to see the video monitor if mounted on the seat-back of a typical First and Business Class seat.

Figure 53 – Seat-Arm Monitors

Therefore, the monitors were installed in flip-out arms that folded under the front of the seat. When the passenger wanted to watch a video, they would press a release button and then reach down and flip-up the monitor and lock it into viewing position.

Unfortunately, there have been cases where the locking mechanism used has not been as robust as the designers imagined, and if bumped, the monitor would sometimes slam down out of position. This could result in damage to the monitor, the arm mechanism, and in some cases injury to the passenger.

There was also the issue with having to have the monitor power itself off when in the stowed position. As these flip-out monitors were stored in small boxes under the front of the seat, if power wasn't removed they would become quite hot and significantly reduce their useful life. Different vendors used different tricks to accomplish this monitor switching on/off: from simple mechanical switches to magnet reed switches to infra-red light bouncing off of some foil tape. Each method has its pros and cons and each vendor selected what they thought was best for their particular installation. In some cases vendors used more than one technique.

With regard to the actual installation, typically, they would be installed in a removable center console, and this console acted like a single Line Replaceable Unit (LRU). If there was a monitor problem, the entire assembly would be removed and replaced. In many instances this was faster than to try and remove/replace a defective video monitor while in the field.

A variant of this monitor installation was also used in some bulkhead seats in Economy Class. Those are the front row of economy where there is no seat in front of you to hang a monitor off of. Sometimes the monitors were installed onto the bulkhead, and sometimes they were installed on flip-out arms.

Different airlines and different aircraft types had appropriate reasons for mounting in one location or the other.

Probably the biggest issue with this mounting location was actually the cable run up through the arm and the "knuckle" joint that let the arm articulate. These are very complex assemblies and provided many areas that could abrade and damage the cabling.

Many newer Business Class seats are now using the old First Class seat design, which I'll be talking about next. And in these new Business Class seats the video monitors are much larger and are attached to seat fixtures/monuments that are fixed. This allows for much larger monitor sizes, with 17" and 19" monitors as fairly common.

Seat-Fixture/Monument Monitors

While primarily intended as a discussion on First Class monitor installations, there has certainly been much movement to install these monitor types into the newer

Business Class seats. As the new business class seats are typically the old First Class seats.

With International Long Haul First Class seats moving toward individual pods, and in some cases individual birthing compartments, the monitor installations have become more complex and the monitor sizes have become much much larger.

Many of the First Class monitors are now 24" or larger. These are typically in fixed locations in a monument in front of you.

Figure 54 – Seat-Fixture/Monument Video Monitors

The only real issue with this type of installation was with the decoding of the audio associated with the video program and where that was occurring.

Maintaining "Lip Sync" is a major issue with IFE systems. When the audio and video are both being decoded within the same unit, there really isn't an issue. However, with different units receiving and decoding the audio and video in different places and at different times, keeping everything in-sync becomes a bit of a chore.

Figure 55 – Monument Video Monitors (United B777 Business Class)

Seat Box

The Seat Electronics Box (SEB) goes by various names depending on the vendor and system being discussed. Names such as: Seat Electronics Box, Digital Seat Electronics Box, Distribution Box, Digital Interface Box, and other similar names are not uncommon. All of these boxes typically perform the same basic functions, they decode audio and video and process command and control signals.

Now, there are several systems on the market that utilize a "smart" video monitor which has the A/V decoding and data processing wholly contained within the display. These systems may or may not utilize a "Seat Box" as a power supply, converting the aircraft 115VAC to the DC voltages required by the monitor. However, generally speaking, there is usually some kind of electronics box in or around the seat that sits between the in-seat electronics and the IFE/CMS distribution system.

Traditionally these boxes contained power suppliers, A/V decoders, telephone interfaces, microprocessors, etc. They were essentially three individual computer systems with several proprietary i/o interfaces all packaged into a single housing.

Seat Box Mounting Locations

Traditionally, the SEB is mounted on the seat leg and is covered by a shroud that prevents the passenger from kicking or damaging the unit.

However, as this intrudes into the passenger's foot space, some alternate locations have been tried, such as under the seat pan.

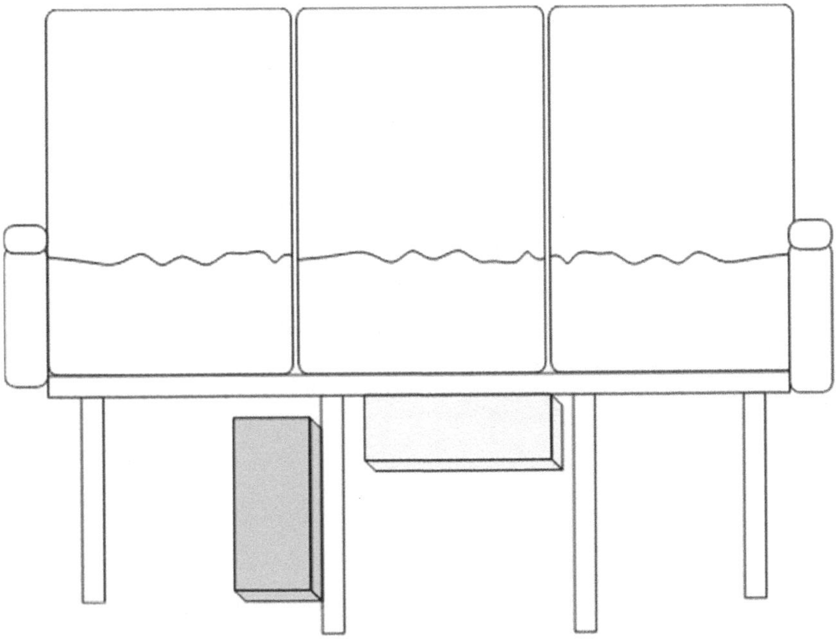

Figure 56 - Seat Boxes mounted on the Seat

While the under seat pan installation has several advantages, some disadvantages such as heat dissipation, angle of airflow, mounting complexity make it a more difficult overall fit.

Also, I have seen seat-leg installations where boxes were mounted on both sides of the same leg, back-to-back on a single mounting plate. This led to the boxes overheating as they were basically pouring heat into each other. So, Heat Dissipation is a key issue with where and how you mount these boxes. Another key issue is how the cabling is routed through and around the seat.

There have also been installations where the Seat Box wasn't mounted on the seat at all. With the nice advantages that mounting the box off of the seat:

- ✈ Reduces seat complexity
- ✈ Reduces seat certification costs
- ✈ Increases passenger comfort
- ✈ Increases system reliability

Things that need to be addressed with mounting the SEBs off of the seats are mostly cable runs, possibly larger cabling and possibly extra shielding. Maintenance access to the boxes also needs to be accounted for. Although SEBs mounted off of the seats typically do not need the extra cost and weight of the shroud and typically have better air-flow and cooling opportunities.

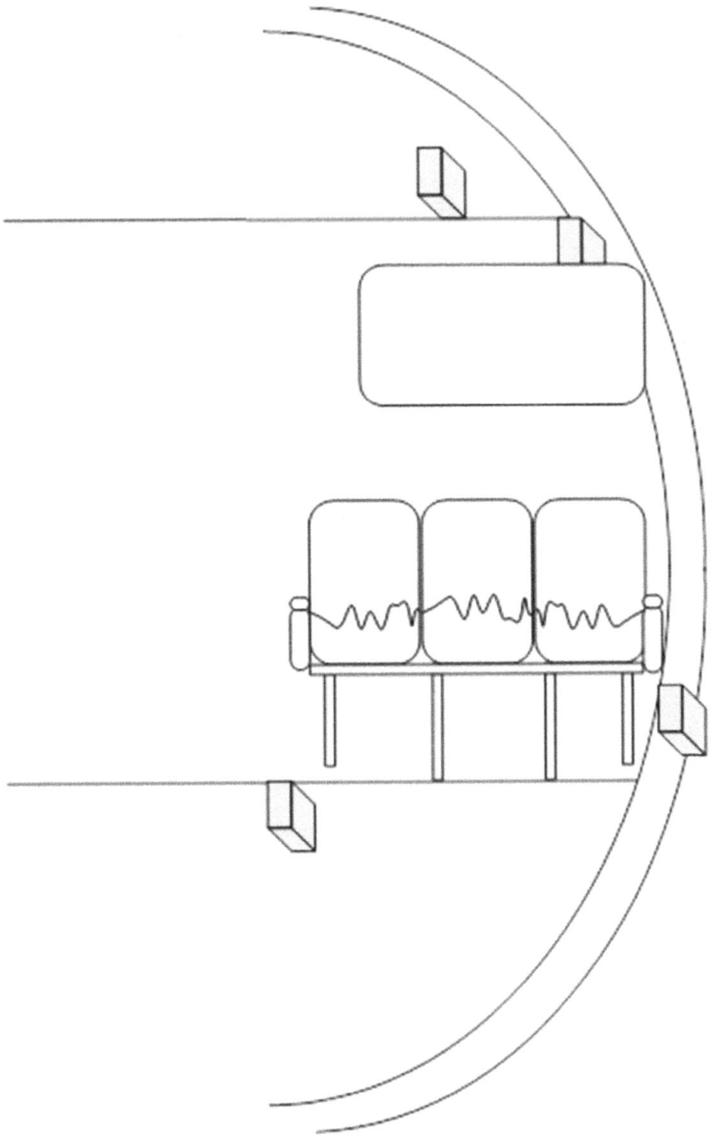

Figure 57 - Seat Boxes mounted off of the Seat

Mounting the SEB off of the seat is typically the best option when it is available to you.

Also, moving the seat box off of the seat allows for more creative use of the available space. Such as designing an

SEB for not only three passengers, but six or nine or more. The limitation really being on the cabling and space available for these non-traditional seat boxes.

However, the down side being that the more seats from one box, the bigger the impact if the box fails. It also increases wiring run length and possible EMI issues. System weight and cost are also impacted but whether that's a good thing or a bad thing depends on the system.

There have been some systems that have tried installing card cages in remote areas (typically under the floor) instead of having dedicated SEBs. In my experience, these have proven to be more problematic than the theoretical benefits gained by fielding such an architecture. Issues typically having to do with: Motherboard stresses, ESD, component fragility, contaminants, socket/pin alignment issues, etc.

Airframe-side Equipment

Head-End or No Head-End?

To have a head-end typically means a dedicated rack for a bunch of equipment in one location (File Server(s), Hard-Disk Arrays, System Control Units, DVD/Blu-Ray players, disk stowage boxes, etc.)

A system without a head-end typically distributes the system/network control functions around to different LRUs and even the source equipment can be located in different areas.

For large aircraft, there is usually a dedicated area for the head-end equipment. In some cases this is a dedicated closet or cabinet called the VCC (Video Control Center) that houses the system control unit, and sometimes data loading devices, with the bulk of the system installed onto pallet above the passenger ceiling panels and the rest in the EE-Bay.

On some aircraft, these systems can weigh over three (3) metric tonnes. (About 6,600 pounds). That weight is usually distributed over as broad an area as possible, with some equipment on the seats and some above the ceiling, some in the sidewalls, and some in the EE-Bay.

For equipment mounted in the EE-Bay, there are issues with long cable runs, typically heavier gauge wiring, complex cable routing issues, etc.

For equipment mounted above the ceiling or in the sidewalls, some issues are noise abatement (do the boxes have noisy fans installed?), and cooling is not as easy as when the boxes are mounted in the EE-Bay.

For units mounted in a dedicated VCC, there may be mounting orientation issues and of course the same noise abatement and cooling concerns as units mounted in the overhead.

Additionally, any units with forced air cooling must have fire detection sensors in the exit channel so as to detect smoke in the case of catastrophic box failure.

Overhead and Bulkhead Video Monitors

Not all aircraft use in-seat video monitors, some use monitors that drop down (rotate down) from the ceiling and/or are fixed onto bulkheads

Figure 58 - Retractable Overhead Video Monitor

Typically, monitors that drop down from the overhead are only used in Narrow-body aircraft, although there are many older Wide-body aircraft that still have them installed.

This is just a follow-on concept to the large cabin video projectors. Only instead of a single projector screen there are smaller monitors distributed throughout the cabin. Although "functionally" they operate the same as a single projector screen, all passengers watch the same movie or TV program or short subject film at the same time. Individual passengers do not have control of the video screens.

Flip-down Monitors

These are monitors that when they are not being used, are tucked up under the baggage bin so that the back of the monitor is flush with the bottom of the bag-bin. And when they are being used, they flip down and present their screens to typically two to four seat rows depending on the size of the monitor.

Probably the biggest danger here is that of "Head-Strike". That is, the monitor striking a passenger on the head, or a standing passenger hitting the monitor. Some other issues are: the noise of the motors may get louder over time, limited viewing angles, brightness/contrast and color differences are exaggerated when passengers can view multiple monitors simultaneously.

There was a time early on when there used to be "fixed" monitors under the bag-bins. But for the most part these have all been retrofit out and are no longer being installed.

When looking at overhead monitor installations the question that always arises is, how many monitors do I need to install? The answer is universally, it depends. The biggest factor in determining how many monitors are needed is how large are they. A rule of thumb we have used for many years is 1" of diagonal is equal to about 1' of viewing distance. So, a 7 inch monitor should be viewable from about 7 feet away.

That distance, when used with and compared to the seat pitch, the distance between seat rows, lets you determine how far apart to place the monitors. This is to ensure that all passengers have a reasonable chance of watching something on a monitor near them.

Also keep in mind that with ceiling and bulkhead mounted monitors, passengers can see a lot of them at once, even if they can't see a clear picture. They will notice things like color balance and this can be a big annoyance to many passengers.

Bulkhead Monitors

Unlike the small in-seat video monitors that may be installed onto a bulkhead for a single passenger, these are large monitors intended for multiple passenger viewing.

These are typically installed high on the bulkhead and directly under the bag-bin.

These units can be recessed into the bulkhead so the face of the monitor is flush with the bulkhead, or they can be surface mounted and sit proud of the wall.

In any case, issues here are typically related to: cooling and air–flow, how the cabling enters and exits the monitor, how the cabling/connectors are held in place, and the overall viewing angles of the monitor.

Monitor Technologies

There are numerous video monitor technologies available from Cathode Ray Tube (CRT), Liquid Crystal Displays (LCD), Plasma and Organic Light Emitting Diode (OLED) to name the most popular.

For the most part, CRT and plasma displays are no longer used. When talking about flip-down and bulkhead monitors we are usually only talking about LCD displays. And while OLEDs have many advantages, they are currently still only used in the smaller in-seat video monitors and touch-screen type Passenger Control Units if they are being used in a system at all.

Antennas

Just a short bit here on antennas. Cabin Entertainment and related systems are increasingly using more and more antennas both inside and outside the skin of the aircraft. From Satellite TV and Satellite Radio systems to voice and data communications and security systems, all vying for precious space to have their antennas installed without interfering with the plethora of antennas used by the flight crew.

Figure 59 – Satellite Antenna Radome

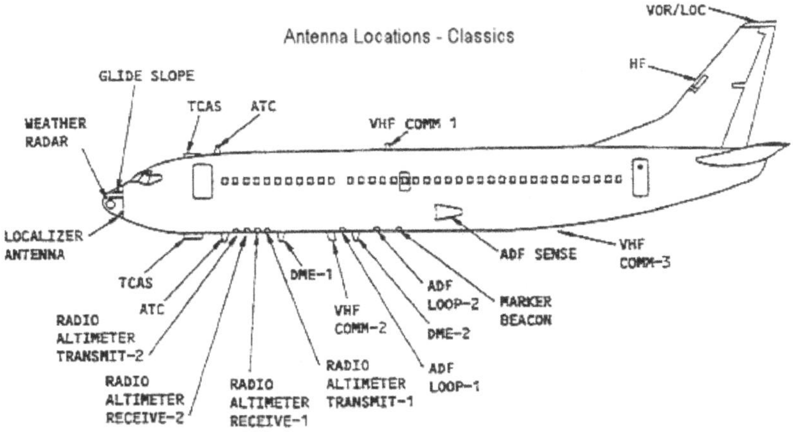

Figure 60 – Older B737 Aircraft Antenna Farm

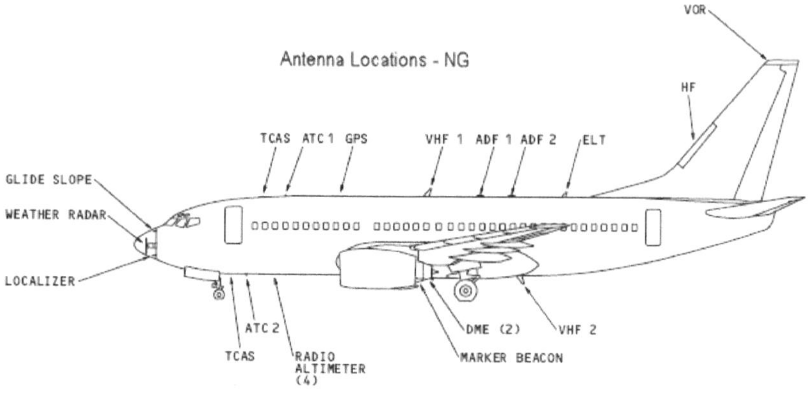

Figure 61 – Newer B737 Aircraft Antenna Farm

There are quite a few issues involved with putting antennas on the outside of the aircraft. Always keeping in mind that SAFETY is the number one priority.

Some of the issues are Drag, Vibration, and Signal Attenuation of the Radome (if installed) and of course the penetration of the pressure vessel. Additionally, there is the issue with potential interference with other antennas and other aircraft systems. These are all critical issues and need to be addressed when looking at antennas mounted onto the outside of the aircraft.

Cabling

Without cabling nothing on the aircraft would work. Yes, I can hear some of you say, wireless!, they could use wireless! Well, for some things yes, but you need to power the millions of bits on the aircraft and that means you need to supply electricity, and that means wires. Miles and miles and miles and miles of wires.

Some small signal wires 32AWG (0.008") to as large as 4/0AWG, which is about as big around as a thumb (0.75").

More typically the electrical wiring falls between 10AWG and 26AWG. With 16AWG, 18AWG, 20AWG and 22AWG being the most common sizes.

Now, that's just for standard wiring, there are also numerous sizes of coax cables, and biaxial and triaxial cables installed, as well as other "special purpose" cables. And let's not forget about fiber optic cables either.

There are several issues that need to be considered here, the most obvious one is that the wire is sized correctly for the amount of current and voltage that it will be expected to carry. So assuming that was done correctly, other things you need to watch for are: cable separation between different types of signals and systems being carried on a particular set of wires, clamping pressure, clamping intervals and the bend radii of the cable.

The wire routing could be an entire book unto itself. Suffice it to say that the installer needs to follow the airworthiness authority guidelines for where the cabling can be run, what it can run next to and what wire separation there needs to be. Additionally, all of the aircraft manufacturers also have guidelines that they use internally to ensure that safe and repeatable installations are properly accomplished.

Now, with regard to cable bend radii, FOLLOW THE CABLE MANUFACTURERS GUIDELINES. Different cabling has different characteristics and can sustain different amounts of bending. When in doubt the following "Rule of Thumb" can be used:

- COAX = 8x Outside Diameter (O.D.) of the Cable
- Non-COAX = 3x O.D. of the Cable

That is, if you are missing the specific cable bend radii information for a particular piece of cable, a typically safe assumption would be that for a 0.25" O.D. coax cable, the MAXIMUM bend radius would be 2.0".

What's wrong with the following picture?

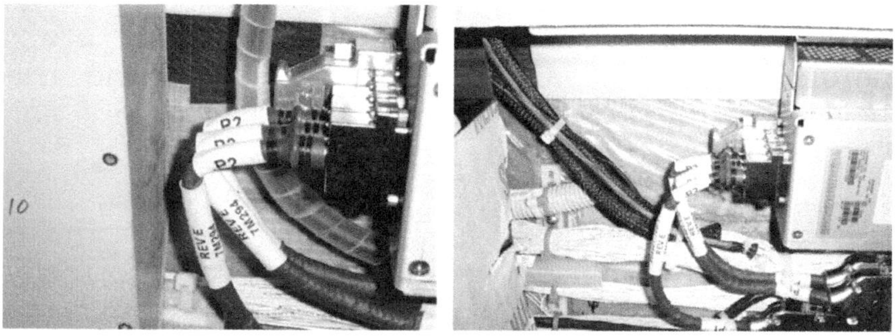

Figure 62 - Cabling, what "not" to do.

Did you notice how the cable is bent back upon itself?

This particular installation issue caused excessive stress on the wires inside the connectors which in turn caused them to only have intermittent connections. This then in turn caused the system to work sometimes and not work other times. Usually it worked while they were trying to find the problem and worked only intermittently in flight. Once the cabling was properly installed, the system worked fine.

One other thing to keep in mind is "Drip Loops". Believe it or not, there is a lot of water in aircraft, mostly due to condensation. You can have a lot of rain above the passenger ceiling and along the sidewalls without the passengers ever realizing it.

Unfortunately, water tends to collect on the cables and flow down along them until it hits a connector. And that is a prime location to get water into the electronics equipment. And as you know, water can let the magic smoke out of the electronics and they stop working. So, we add drip loops to the cabling at the point where it enters the box.

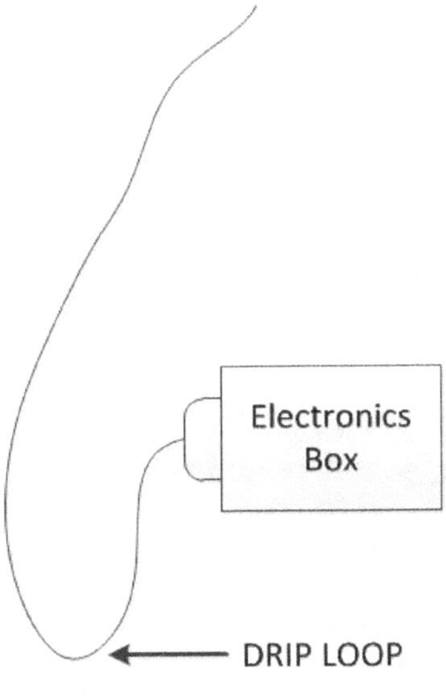

Figure 63 – Drip Loop

As you can see in the drawing above, as water flows down the cable, it would fall off the cable at the bottom of the drip loop instead of flowing into the connector. This then prevents the water from entering the LRU at the connector. Yes, you could use hermetically sealed connectors but they are very costly, especially when compared to the negligible cost of adding a simple drip loop.

This is the same reason the boxes are typically never installed with the connectors facing upward from the box. They are nearly always installed with the connectors facing the sides or downward.

So, we talked a bit about the cabling behind the sidewalls and in the ceilings, but how do you protect it from the passengers in the cabin?

Seat-Track Covers

Traditionally, cabling runs along the seat track. The Seat Track is an aluminum channel that the seats are bolted to on the floor; this in turn is covered by a plastic Seat Track Cover. Then, the cabling runs in a Cable Raceway which sits under the seat track cover.

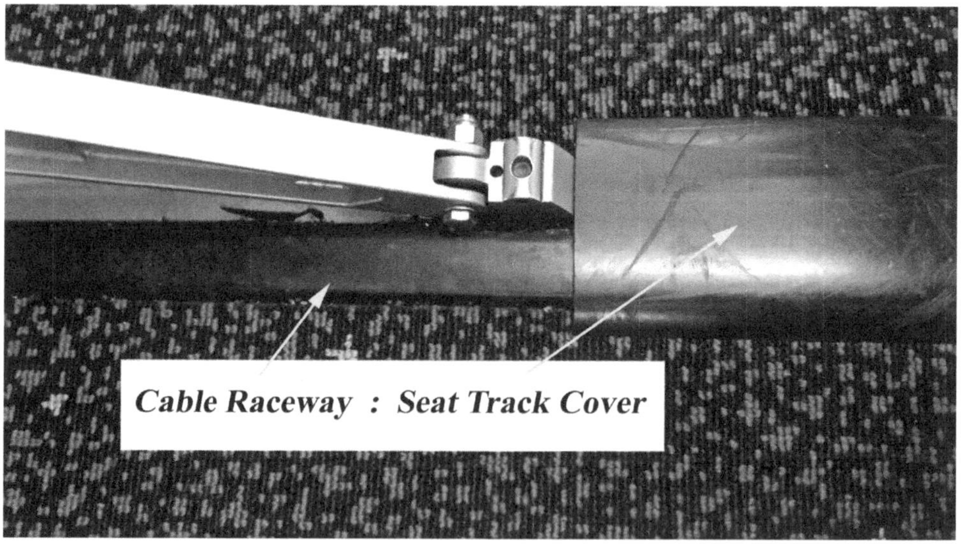

Figure 64 - Seat Track Cover & Cable Raceway

Typically all of the cabling runs along one side of the seat track, as illustrated in the figure above. Also, there can be one or more cable raceways under the same seat track cover, it just depends on the width of each item.

Sometimes there is just so much equipment in the seats requiring a multitude of cables, and differing power requirements, that the cablings is run on both sides of the seat track. This makes for a very wide (could be over 8" wide) and none-to-attractive seat track cover stretching along the floor between the seats. Additionally, there have been cases where cabling has been run along both seat tracks in order to maintain proper cable separation and to prevent interference from different systems installed in the seat.

Probably the biggest issue here is the lack of proper installation oversight and improper maintenance practices. That is, the cabling is being installed under the seat track cover, but NOT within a proper cable raceway. A cable that is just tucked under the seat track cover will eventually vibrate out from under it and be exposed to passengers.

Couple a small exposed wire with the sharp edge and concentrated pressure of a woman's high-heel shoe and it's a recipe for disaster. I have heard that this exact situation occurred on at least one flight and caused arcing and sparking. Not a good thing at all. And then there is also the possibility that it could be a trip hazard in an emergency.

Okay, we've covered a fair amount of the basics, now on to a little bit more detail.

Systems

When an airline is contemplating a new system, they need to look at which aircraft it will be installed on, existing IFE in their fleet, will it be installed as a retrofit installation, or as part of a new aircraft purchase. And of course onto their wide-body or narrow-body fleets or both.

Retrofit Aircraft

Retrofitting systems onto existing aircraft is by far the largest market for vendors and can offer the fastest return on their investments. This segment is also where the most innovation seems to take place, as the aircraft manufactures are generally very risk adverse and are not prone to taking chances on new technologies.

On the positive side:

- Typically constraints on the vendors are few
- The only regulatory issues are related to safety
- For the most part, airlines can get what they want, when they want it
- The airlines typically provide their own labor to accomplish the installation
- Installations on multiple aircraft can take place daily or weekly depending on the system and the ability to quickly install it
- Can utilize Just-in-Time delivery scheduling to the installation sites

On the negative side:

- The vendors are expected to provide the installation engineering
- Vendors are typically asked to provide an already approved Supplemental Type Certificate (STC)

    - If no STC exists, then a Project must be opened with the FAA, or other equivalent organization outside the US, and a lengthy approval process is begun.
- The vendors typically need to have personnel on-site to oversee the installation
- Finding optimal locations for boxes and cable runs may be a challenge.
- The airline may ask the supplier to provide the installation labor.

The only real "gotcha" here is when you design an installation package and the airline changes the initial aircraft at the last minute. Unfortunately, this happens way more often than it should.

The ripple on effect of what too many in the airline consider a simple change, can more often than not triple the amount of work that needs to take place.

If you start a project with Tail # 1234, and the airline switches at the last minute to Tail # 2278, then all of the governing paperwork needs to change, the drawings reviewed and updated for changes to circuit breakers, wiring runs, equipment locations, electrical load analysis, weight and balance, etc., etc., etc. This has an enormous impact to the vendor and needs to be contractually accounted for.

With regard to schedule, the equipment lead-time can have a negative effect on the program if the airline doesn't release purchase orders in a timely fashion.

In general, there is a lot of flexibility with retrofit systems and airlines are more likely to get what they want in the retrofit market than from a new aircraft delivery.

New Production Aircraft

This is where the IFE vendors provide equipment and systems to the aircraft manufacturers (OEMs).

On the positive side:

- The vendors have a more stable income stream if their equipment is baseline or a preferred option at the OEM
- The OEM takes care of all the certification tasks such as installation engineering and providing the labor

On the negative side:

- There are typically many system constraints imposed on the vendors by the OEMs
  - Size, Shape & Mounting methods of boxes
  - Number and type of boxes
  - Number and types of cable bundles
  - Cable routing and disconnect points
  - Connector types and pinouts
  - System LRU on dock dates are typically 4 to 5 times shorter than a typical retrofit installation
- Vendors have to deliver multiple shipsets of units many months, sometimes as much as a year, before actually getting paid for them
- Airlines typically pay for deviations to OEM "baseline" system configurations
  - This is different than an "options" catalog which simply lists pre-approved options to the baseline system.
  - If an airline wants something not already approved by the OEM, they are typically asked to pay Millions for any changes.

Probably the biggest hindrance here is the inability of most OEMs to be flexible. This isn't "good" or "bad", it just is.

Aircraft OEMs want to build every aircraft the exact same way to keep their costs down and production yield up. Any changes to these baselines cause production headaches and possible errors to inject themselves on the production line.

Embedded IFE System Vendors

Okay, so now on to some of the system suppliers. But first a HUGE DISCLAIMER: what follows is NOT a comprehensive list of all IFE vendors. New or emerging vendors pop up every year. Some survive in niche markets, others are absorbed by the big guys, and others simply fail. I have seen some great technologies introduced over the years that would save the airlines billions of dollars but failed because the vendors were not big enough to play in this market.

The current major players in the IFE market are:

- Thales
- Panasonic Avionics (Matsushita)

Other companies in this space are:

- Lumexis
- Rockwell Collins

The above all offer traditional "wired" type fixed installations.

However, the explosion of cheap tablet computers in the consumer electronics arena has caused numerous companies to appear overnight trying to sell these portable tablets into the aircraft IFE market.

Additionally, every year at the Airline Passenger Experience Conference (APEX), (formerly called the WAEA (World Airline Entertainment Association)) conferences there are at least 3 or 4 new vendors showing their wares. That's EVERY YEAR. Unfortunately, their success rate is not very good.

Having a killer application or technology is not enough. The Product Support and Maintenance side of the business is just as important if not more so...

Before an airline spends $100MM+ to equip its fleet with new IFE, they want to ensure the vendor is going to be around for a while. And, that they know how, and are capable of, supporting their products globally.

Wide-Body Aircraft

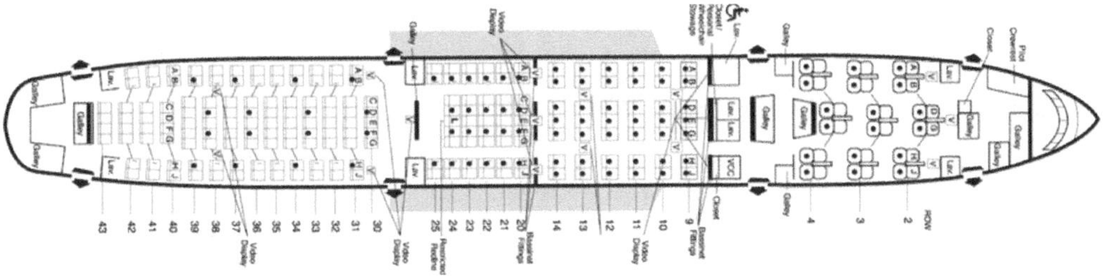

Figure 65 - Wide-Body LOPA (B777)

These aircraft are defined as: Twin Aisle, with multiple classes, multiple zones, and different levels of service in each zone. Even through aircraft such as the A380 and B747 have multiple levels, or decks, they are still just wide-body aircraft, with a lot more capacity.

These aircraft are typically long-haul aircraft with typical flying times of greater than 8 hours.

While traditionally the First Class cabins have had sleeper seats for some time, several airlines are moving to an almost "Super First Class" cabin area with individual sleeping compartments that are more reminiscent of what you may see in a luxury train cabin.

Also, with the amount of electronics in some of the F/C areas, some airlines have taken to installing false floors for the seats to be installed upon that then also gives them unobtrusive places to route the cabling, without having to utilize seat tracks and seat track covers that may mar the appearance or aesthetic of the First Class cabin.

As you can well imagine these systems tend to have the most capability and by far the largest footprint, weight and cost of any other aircraft type for its IFE system.

As previously stated, the approximate weight of a full up AVOD system is in the neighborhood of 4,400 to 6,600 pounds (2 to 3 metric tonnes). With a price tag that can be as high as $7MM US Dollars per aircraft.

Wide-Body Embedded IFE Suppliers

The big two suppliers in this market are:

- Panasonic Avionics Systems
- Thales

Panasonic Avionics (Matsushita)

Panasonic has a huge range of products and systems for this market. Generally speaking, they are all Ethernet based and can provide various levels of functionality from simple broadcast audio and video to complex AVOD with interactive gaming, shopping, off aircraft communications and other similar applications.

Figure 66 - Panasonic Wide-Body

Figure 67 - United B777 Business Class (Panasonic)

Thales

What I just said for Panasonic also holds true for Thales. They essentially have the same types of product offerings. The physical implementation of each of their systems is where they differ. But they both offer complex AVOD and data driven systems.

Figure 68 - Thales Wide-Body Head-End

The figure above shows a rack of Thales equipment, taking up 70 MCU of space that can provide AVOD for up to 550 passengers.

These types of systems utilize multiple cable runs, special QoS (Quality of Service) protocols and lots of computing power at each seat.

Figure 69 - Thales B767

The figure below is that of a typical in-seat cable interconnect. For the seat vendors, they hate these interactive AVOD systems as they have miles of cabling that ends up being run throughout all the aircraft seats. This is by no means restricted to Thales. Panasonic has exactly the same issues.

3GCN/3GCN-W Seat Harness Schematic Diagram

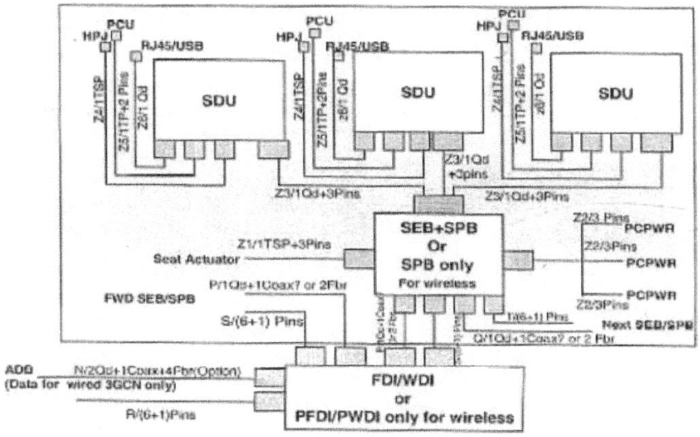

Figure 70 - Thales Seat Harness Schematic

One of the <u>great</u> things that these vendors do is to set up full scale test labs with an entire shipset worth of hardware.

This allows them to test all the equipment, and especially the software, at a scale that mirrors what will be on the aircraft. As shown in the figure below, hundreds of displays, seat boxes, PCUs and the associated cabling and head-end units are all set up and loaded so the developers, and the airline personnel, can see what it is they are getting. From the latency of pressing a call light button to any issues with lip-sync, these are all identified on these test rigs.

Figure 71 - Thales Test Lab

Narrow-Body Aircraft

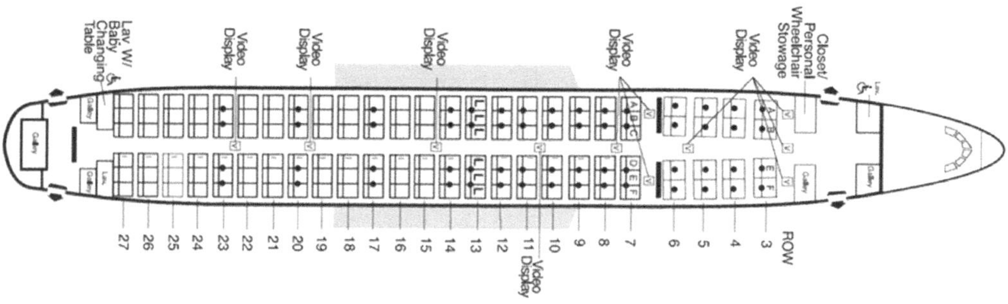

Figure 72 - Narrow-Body LOPA

These aircraft tend to be Single Aisle, One or Two classes, multiple zones, with different levels of service in each Class.

They are typically short haul, less than 4 hour flights, with the average probably closer to 2 to 3 hours.

These aircraft don't have nearly as much space available. Where are you going to install all of the IFE equipment?

Regional Aircraft "could" also fall into this category as many regional carries are now also installing IFE systems in order to remain competitive.

Narrow-Body Embedded IFE Suppliers

The big two suppliers in this market are:

- Panasonic Avionics Systems
- Thales

With Lumexis taking up a distant but growing third place.

The list should look very similar to the Wide-Body list. Mostly the same players, with the same markets. But in most cases they have very different equipment with different system capabilities.

Panasonic Avionics (Matsushita)

Again, their systems are primarily Ethernet based and can provide a full range of optional capabilities. From simple A/V to complex interactive functionality.

Thales

Same as above for Panasonic. Similar capabilities just different implementations of technology.

Figure 73 - Thales A321

Lumexis

Now, here we change things up a bit with Lumexis.

Rather than using a copper Ethernet based system, Lumexis uses a full implementation of fiber optic technologies.

Additionally, whereas the other vendors offer some simple systems with basic A/V functionality, Lumexis offers only a full AVOD system with varying degrees of optional interactivity.

Figure 74 – Lumexis Seatback Video System

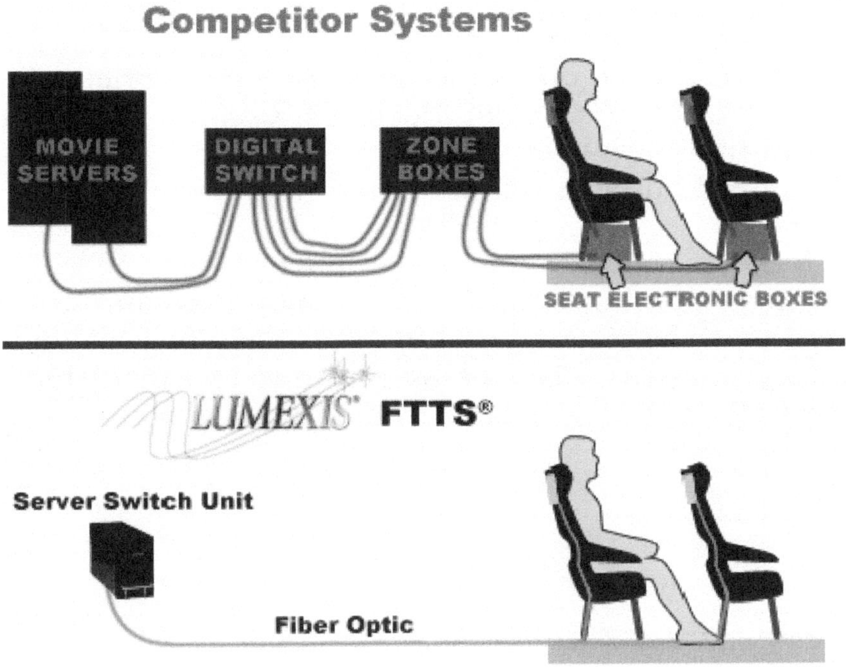

Figure 75 - Lumexis IFE Comparison

Here is an example from an early trial of the Lumexis system for which published data is available:

- US Airways 3 Month Flight Trial RESULTS
- Entry to Service : 9 March 2009
- Total flights @ 4/day: Nearly 300 Flights
- Total flight time @15hrs/day: 700+ Hours
- Total operating hours: Over 1,000 Hours
- Entertainment usage by passengers:
    - 40,210 videos watched
    - 21,156 preview videos watched
    - 18,305 games played
    - 224,872 music tracks started
- System re-boots during Flight Trial: ZERO
- LRU failures: 1 Video Display Unit

That is a phenomenal reliability number for a first entry into service. Especially with regard to the zero re-boots of the software.

Now, before I move on to the next section I'd briefly like to mention Rockwell Collins. Rockwell Collins was a huge player in the IFE market and then after a series of missteps they dropped out as a contender for quite a while.

They have recently revamped their narrow-body offering and have provided a system with HD monitors and Blu-Ray players to at least one aircraft. We'll have to wait and see how they progress. Perhaps by the time you read this they will be well on their way back from IFE obscurity.

Portable Electronic Devices (PED)

These devices are characterized by not being permanently connected to the aircraft for power or data.

- Typical Passenger carry-on devices
    - Nintendo DS type devices
    - Laptop, Tablet, Netbook Computers
    - DVD Players
    - MP3 Players (iPods)
    - Virtually any electronic device that can be used in a portable fashion
- Airline Provided PED
    - Many different types and styles
    - Mostly similar to tablets or tablet like devices

Figure 76 - Nintendo 3DS

Federal Airworthiness Regulations (FAR)

You may not realize it but PEDs are actually regulated. The relevant FAR is shown below. (Please note that the emphasis on certain words and phrases is mine.)

Sec. 91.21 - Portable electronic devices.

(a) Except as provided in paragraph (b) of this section, no person may operate, nor may any operator or pilot in command of an aircraft allow the operation of, any portable electronic device on any of the following U.S.-registered civil aircraft:

(1) Aircraft operated by a holder of an air carrier operating certificate or an operating certificate; or
(2) Any other aircraft while it is operated under IFR.
(b) Paragraph (a) of this section does not apply to --
(1) Portable voice recorders;
(2) Hearing aids;
(3) Heart pacemakers;
(4) Electric shavers; or

(5) Any other portable electronic device that the operator of the aircraft has determined will not cause interference with the navigation or communication system of the aircraft on which it is to be used.

(c) In the case of an aircraft operated by a holder of an air carrier operating certificate or an operating certificate, the determination required by paragraph (b)(5) of this section shall be made by that operator of the aircraft on which the particular device is to be used. In the case of other aircraft, the determination may be made by the pilot in command or other operator of the aircraft.

[Doc. No. FAA–1998–4954, 64 FR 1080, Jan. 7, 1999]

Reference Advisory Circular 91.21-1A dated 2OCT00. Also applies to section 121.306, 125.204 and 135.144 aircraft.

So, I highlighted paragraph (5) because that's what lets you use your iPad or Laptop or whatever on the aircraft. It's the "Operator of the Aircraft" that is giving you permission to use your electronic device. If at any time the aircraft operator, or pilot, determines that electronic devices in the cabin may be causing a problem, they can immediately order all items shut off. Non-compliance by the passenger is a US Federal offense and can get you arrested and heavily fined. This has happened before and most assuredly will happen again.

PED Market

With regard to the airlines purchasing PEDs to hand out for free and/or rent to their passengers, this is growing to be a significant market.

- Market estimate is for portable IFE devices to eventually represent at least 5% of the overall IFE (Hardware) dollars spent
- That translates to approximately $75M USD in 2012 and to over $200M by 2020

Portable IFE

There are far too many entrants in this category to even begin to list here so I'll just start by mentioning the grand-daddy in this market: the digEplayer.

This company basically invented this category and has been through many buyouts and name changes over the years, the latest company to own them as I write this is Total Aviation Solutions (TAS), but they are still using the digEcor company brand name.

- digEcor
    - digEplayer
    - Content:
        - Magazines, Books & Newspapers
        - TV Shows
        - Video Games
        - Movies

Figure 77 - digEplayer

And then of course we have the iPod. Yes, the iPod, I'll get to the iPad in a minute.

The big IFE companies even provide a port that you can plug your iPod into and control it via the IFE system. Below if a picture of the Panasonic eXport interface.

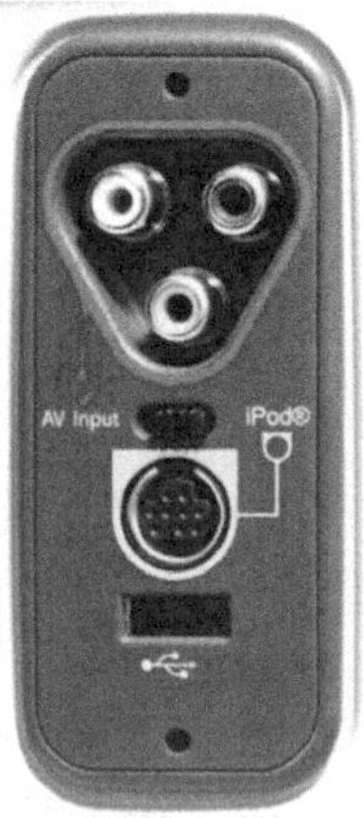

Figure 78 – iPod Port

This then leads us to mention the iPad. (Or any other tablet such as the Android or Windows based units out there.) It seems like everybody and their brother has set up a company to sell tablets into the airlines.

What nearly all of these companies fail to understand though are the logistical issues around fielding portable devices. For example, where and how do you charge their batteries? And then once in the field, how do you support failures? How do you contain a Li-Ion battery fire?

And equally important, (okay, not equal to the bit about fire because safety should always be the #1 priority), but how do you get the CONTENT that is required to make them useful? Content that constantly needs to be refreshed, securely transported and loaded, and how do you charge the passenger or airline for it?

I realize that many of these questions have answers and several companies have products in this field. What I'm talking about here are the numerous start-ups that have no clue about supporting products in the airline industry.

These questions, and many many others that are similar in scope, are what typically dooms these start-ups to failure.

Wireless IFE System Vendors

Same disclaimers here about this not being a comprehensive list as there are a lot of players in the market. I'll just name a few to give you a flavor of this market space. In no particular order they are:

- GoGo
  - GoGo Vision which can be a stand-alone product or used with their Terrestrial Broadband system
- Panasonic Avionics
  - They have several Cabin Wi-Fi solution as part of their embedded offering and as part of their Satellite Connectivity option.
- KID-Systeme
  - Their SKYfi system is being trialed on a few carriers
- Zodiac
  - They have the RAVE Wireless solution which can be stand-alone or part of their RAVE embedded solution.
- Global Eagle Entertainment
  - Provides IFE software solutions and connectivity via Ku-Band satellite

Wireless-IFE to passenger PEDs, which I talked about earlier in this book, is certainly doing brisk business.

Below is in image of the GoGo Vision Movie & TV Menu.

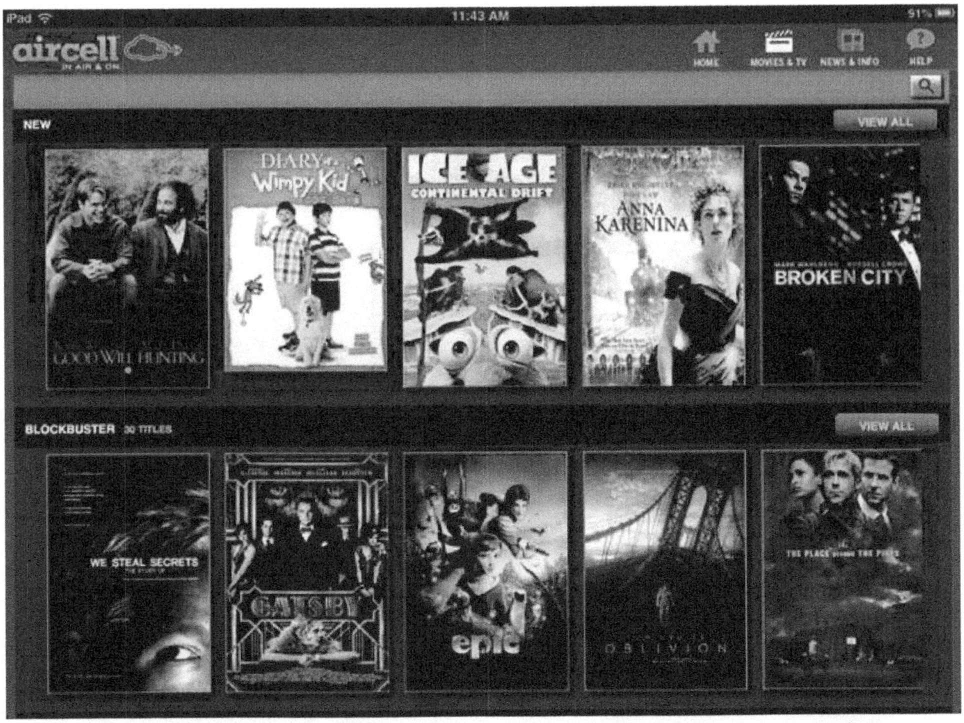

Figure 79 – GoGo Vision - Movie & TV Menu

Programming and Content

Okay, so now I've covered a bit about the hardware that's installed, or carried on-board, now let's chat a bit about the Content and Programming side of things.

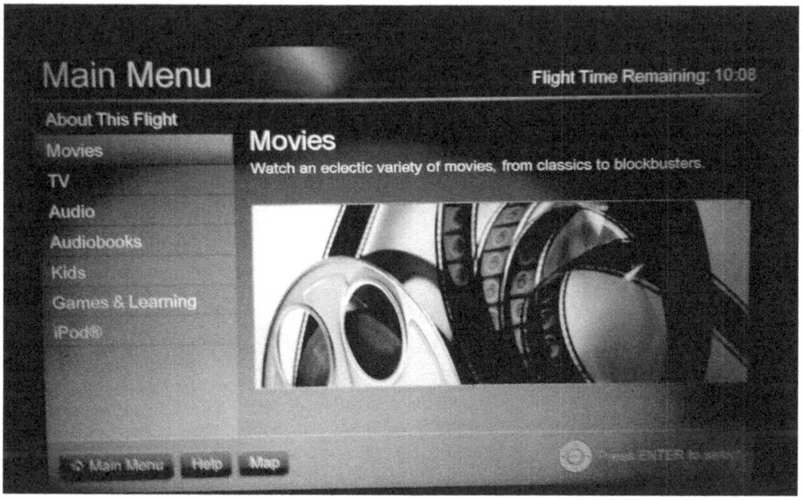

Figure 80 – United B777 IFE Main Menu

When there was only one big projection screen in the cabin, or even a lot of drop down monitors, everyone watched the same thing at the same time. So, the studios developed a financial model that typically charged the airlines a fee for the movie based on:

- Load Factor
    - How many seats were typically sold
- Number of Flights per Month
- And a few other minor variables

Typically they would purchase two movies per month, one for Eastbound/Northbound flights and one for Westbound/Southbound flights.

Then along comes Multi-Channel Broadcast Video and the airlines now have to buy 10-16 movies per month instead of only two.

Then along comes AVOD and now the airlines have hundreds of movies on-board. This became a significant cost increase to the airlines.

But then on the flip side, a cost reduction came when most airlines gutted their IFE departments and eliminated staff and facilities.

You see, when there was only one big screen and everyone had to watch the same movie, the airlines had a staff of people, usually a small staff but a staff nonetheless, that was tasked with editing out all the objectionable bits that the more sensitive fliers may object to.

With the introduction of AVOD, and passengers being able to watch what they want, all of that costly editing came to a screeching halt. If you don't want to watch a movie with a little skin shown, watch something else. If you object to movies with blood splatter and gore, it's simple, just watch something else.

And on a side note, the regional differences around the world were always interesting to note. In the Scandinavian countries for example nudity was tolerated but blood and gore and violence were heavily edited out. While in the Middle East it was just the opposite, no holding hands or kissing was allowed but blood squirting from an open wound or head being chopped off is shown all the time.

Some of the big questions about the content are:

- Who pays for it?
  - The passenger?
    - As they watch/use the service?

        - As part of their ticket?
        - A la carte or per flight?
    - The airline?
        - No direct charge to the passenger
    - Advertisers?
        - Traditional commercial ads
        - Non-traditional sponsorship
- Where does the airline get content from?
    - Directly from studios?
    - From a content aggregator?
    - From the IFE Vendor?
- Editing
    - For smaller screen sizes?
    - For different screen aspect ratios
        - 4:3, 16:9, 16:10, etc.
    - For content?
        - Still happens with some airlines
        - And please keep in mind that many directors and actors won't allow their movies to be edited for content

Pre-Release (Early-Window) Movies

In the US movie market, there are typically six release stages that a movie goes through. Starting with the most financially lucrative to the least, with the exception of the US Military market.

- Movie Theatres
- US Military
- Aircraft
- Cable TV
- Rental
- Broadcast TV

The US military tends to get the movies second as one of only a very few perks that the military has. Most movie studios release to the military second as a way of showing support for the troops. There is certainly no law that mandates this, just a long held tradition of thanks.

Also, airlines don't usually get movies when they are released in theatres because airlines are notoriously lax in their security measures when it comes to safeguarding movie content. That's why the IFE companies have recently gone to great lengths to show the movie studios that their systems are secure and the loading of content can be accomplished securely and with minimal human intervention.

This is an enormous issue when we think about airlines providing A/V content to passenger's own carry-on devices. For the savvy computer geek, or traditional hacker, there is virtually no content that they can't capture and copy once it hits their computer. Less so with tablets, but ripe for the taking with just about any laptop loaded with the right software tools.

Now, let's move on to Satellite TV for a moment. (Real Satellite TV, not TV over a satellite internet connection.) Many people seem to think that once they install a Satellite TV system that they can just show the passengers whatever they want. That is simply not true. The user needs a specific LICENSE to show this content to its passengers. Satellite TV hardware is only ½ of what you really need:

- Licensed Multi-Region TV Coverage
- Content Licensing
- Established relationships with content providers
- Networks typically do NOT own the rights to the shows they broadcast
    - They license the rights to broadcast a particular show in a particular region

You cannot legally just install a satellite TV system and start showing ESPN or CNN to your passengers. You need a specific license from the show owners, not necessarily the network, in order to make shows, carried on these networks, available to the airline passengers.

Social Interaction

Virgin America has an interesting twist on the passenger/airline communications paradigm. As part of their "Red" system, they let you interact with the airline in real time. They also pull your last few posts from any social network they can find you on (Facebook, Twitter, Etc.), and presents your flight history and sends you targeted messages about your final destination.

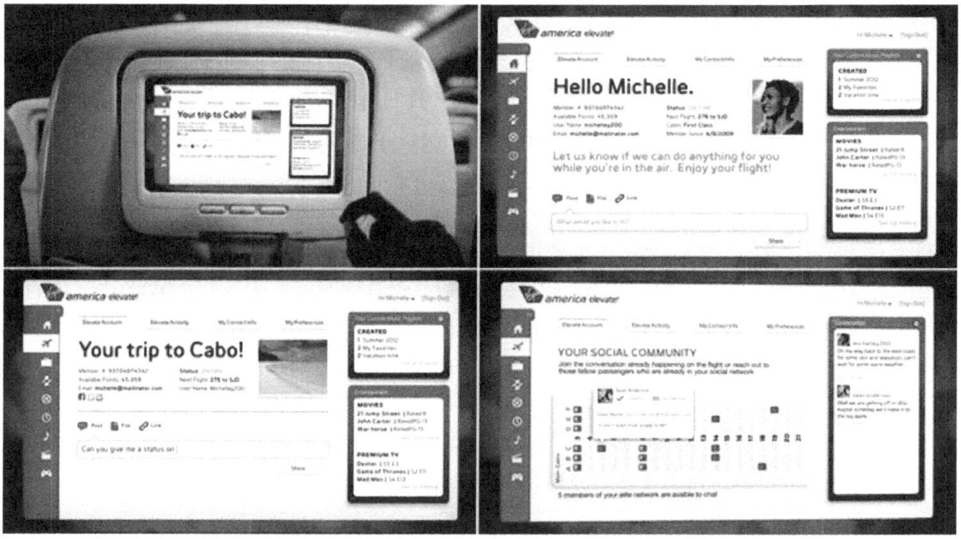

Figure 81 – Social Interaction

They also let you buy drinks for other passengers and give you the ability to text them as well...

It's a wait-and-see right now as to how many other carriers follow suit...

Figure 82 – Social Interaction

Regulatory and Certification Issues

General Guidelines: FAA/JAA/CTA/Etc.

- SAFETY IS THE #1 PRIORITY
- Can the proposed IFE system interfere with flight critical systems?
- Can the system cause a fire or produce toxic gasses if it does catch fire?
- Can it self-extinguish?
- Can it be turned off from a convenient location?
    - Circuit breakers SHALL NOT be used as on/off switches.

The key thing here is SAFETY. None of these systems help the aircraft stay in the air. They are not flight control systems. They are toys to keep the passengers amused during their journey. They are all NON-ESSENTIAL equipment.

Also keep in mind that there MUST be a single power on/off switch that can remove power to the system in the case of an emergency. A software controlled button on a touch-screen panel is NOT sufficient. It could be that the unit the crew is trying to remove power from is the one with the touch-screen and it's the one that is smoking. *That is actually a real scenario that has happened in the past.*

Additional things to watch for in this space are:

- Any specific IFE related Advisory Circulars or Directives?
- Someone in the Avionics group should be checking on this periodically.
- Nobody should be caught off guard by a new regulation or guideline.

Industry Guidelines: RTCA/DO-160

This document contains the electrical and environmental tests that should be conducted in order to prove that a piece of equipment is airworthy. Depending on the function and installation location of the equipment or system, to be tested, the Unit Under Test (UUT) will undergo significantly different testing than another piece of equipment with different functions and installation locations. This document can be purchased by contacting www.rtca.org.

- As I write this, revision level "G" is the latest version
  - RTCA/DO-160G
- It lists qualification testing requirements and tests that should be run on the equipment.
- It lists categories of tests and various levels within each test.
- Specific tests depend on equipment type, function and location.
- Some tests are performed and others are not.
- Some tests are performed on some equipment but with different PASS criteria than for other equipment.

It is critically important to understand the classification of equipment and the specific criteria that units were tested under.

Two vendors can both honestly say that their units pass DO-160 and yet have completely different performance characteristics.

Type Verification could also be via analysis, but testing is required in most cases. There can also be Verification via similarity to other units already certified, EXAMPLE PCU color is different but physically, functionally and electrically it's the same as a previously certified unit could be claimed to have passed all the tests via a "Similarity" statement.

Additionally, airlines should ask the vendors the right questions. I was once at a trade-show and heard an airline person ask a vendor if the unit had been tested to DO-160, to which the vendor happily exclaimed, "Of Course!" And they then went on with their conversation with the salesman extolling the virtues of their equipment.

Unfortunately, I happen to know that that particular vendor had "tested" to DO-160 but they had in fact FAILED several of the tests. But, whether they had passed the testing, or even what testing was accomplished, wasn't the question that was asked. It was only asked if they had "tested" to DO-160. To which that salesman had honestly replied, yes.

Airlines need to ask vendors the right questions.

SEAT Certifications

This topic could also fill a book by itself. So what I'll strive for here is just a bit of an overview.

Seats get installed at the aircraft OEM by the authority of a Technical Standards Order (TSO). This installation certification not only includes the seats but also all the ancillary bits, typically IFE, that are attached to the seat. This does not certify the IFE. It certifies that the seat can be installed into the aircraft with a particular IFE kit installed onto it.

Historically, the seat vendors have been able to simply certify a mass with particular attach points and then could install any IFE equipment that was below that mass and had the same attach points. Over the past several years though as the IFE systems and their wiring has become more complex, the FAA has required that the Seat be certified with a specific set of IFE part numbers. Now, this has made it extremely costly for the airlines when they wanted to upgrade the IFE as it would necessitate a whole new seat certification program and the application for a Supplemental Type Certificate (STC) for the seat.

Several FAA and industry committees have been formed to look at this cost and safety issue and hopefully by the time you read this it will have been sorted out.

To quickly summarize:

- Seats get a TSO on a new aircraft while having installed a particular IFE build.
    - There are FAA and Industry committees investigating the high costs of doing this all the time.
- Seats also get STCed regularly when the IFE changes in retrofit situations.
    - The high costs here are also being researched.
- SEAT Issues: (Example)
    - Seat TSO certified without in-seat video
    - Airline later adds in-seat video
    - Seat impact:
        - Center of gravity changed
        - 16G survivability impacted
        - Possible Head-Strike Issues (Head Impact Criteria (HIC))
        - Cable routing and interference issues

Airframe TSO/STC Issues:

As airlines add equipment that touches the actual airframe, they then have their own STC issues.

- Satellite TV Antennas
- Structural mounting of overhead monitors
- Structural mounting of racks for LRUs

Anything that impacts the basic structure of the aircraft needs its own analysis and substantiation.

Documented Substantiation of Findings

And speaking of substantiation, there is a mountain of paperwork that is required to prove to the certification authorities that what you are doing to the aircraft won't hurt it.

A few of the major reports that need to be done for a system installation:

- Safety Analysis
- Structural Analysis
- Electrical Load Analysis
- Weight and Balance Report
- Failure Modes & Effects Analysis (FMEA)

There have been cases where IFE systems were installed that consumed more than 100KVA, but the airline only had 90KVA ground power carts. So when they wanted to run the system on the ground, for BITE/troubleshooting, they had to start an engine or the APU. This was a very costly lesson the airline learned after the kit was installed. A very costly lesson indeed.

The Electrical Load Analysis (ELA) tells the airline many things about the system. It also lets you look for incompatible equipment powered on the same power bus. For example, having large inductive motors on the same power line as the IFE can cause numerous issues including brown-outs and large voltage spikes on the power line. These in turn can cause anomalies in the IFE system.

Industry Guidelines: ARINC 628

This is a GUIDELINE and NOT a regulation.

Airlines can almost always get better performance and lower costs by NOT adhering to this standard. But then they LOSE the ability to easily change vendors and have common mechanical and electrical interfaces. There is always a trade-off...

ARINC 628 is a multi-part document with each part having a specific focus. What follows is a brief overview of each part of the document. The full document can be purchased by contacting: http://store.aviation-ia.com/cf/store/ .

Part 0 – Cabin Management and Entertainment System – Overview

- General System Architectural Philosophy
- Guidance for the rest of the specification

Part 1 – Cabin Management and Entertainment System – Peripherals

- Defines input interfaces to head-end devices such as Video Players, Cameras, etc.
- Environmental conditions, mounting considerations, noise limits, etc.

This part defines the INTERFACES to the equipment, but not how the units work internally. You will see here mounting envelopes and methods defined but not specific box shapes. If software is necessary for the interface then the control words will be defined. But the top level Operating System such as Linux, or Windows, or some other OS will NOT be defined.

This part also discusses the amount of fan noise the unit is allowed to generate depending on where in the aircraft the unit is physically installed.

Part 2 – Cabin Management and Entertainment System – Seat Interfaces

This Part is about how to attach units to the seat and the internal seat cable routing guidelines.

- Seat mounted equipment considerations
- Passenger Control Units, Headphone Jacks, Seat Actuators, etc.
- Cord Reels
- Armrest mounted Monitor Deployment

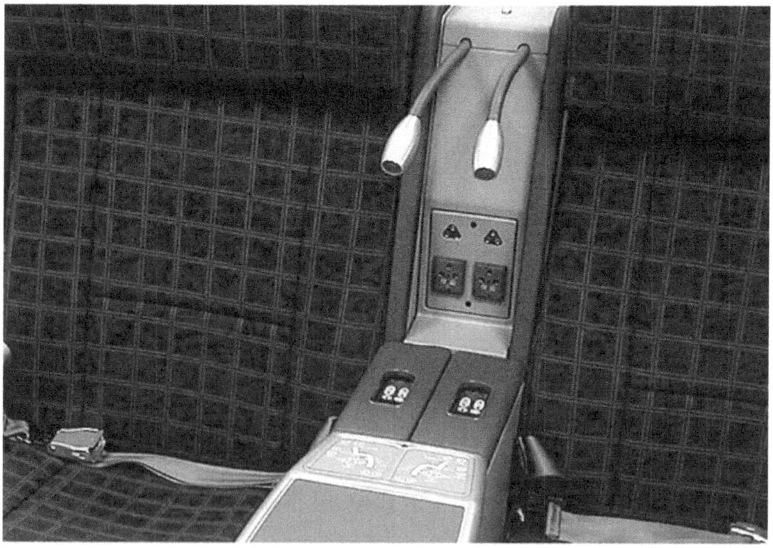

Figure 83 - Seat Equipment

Figure 84 - Seat Equipment – PCUs & In-Seat Monitors

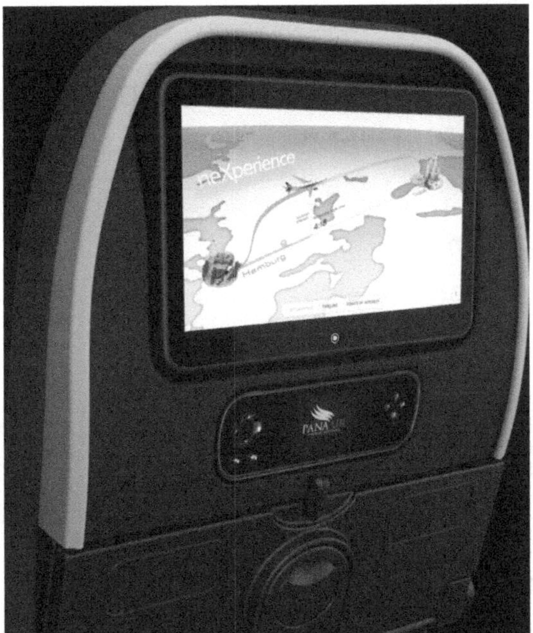

Figure 85 - Seat Equipment – PCU & In-Seat Monitor

Figure 86 - Seat Equipment - Handsets

Part 3 – In-Flight Entertainment System (IFES) to Aircraft System Interfaces

This Part is about how to interface the IFE to different, non-IFE, aircraft systems.

- RS485 protocol
  - Command and control
- ARINC 429
  - IFE uses READ ONLY access
- PA Interfaces
  - Audio from the PA and PA keylines
- Decompression
- Built-In Test Equipment (BITE)
- Printer Sharing
- Aircraft Communications Addressing and Reporting System (ACARS)

Figure 87 - ACARS

Part 4A – Cabin Management and Entertainment System – Cabin Distribution System - Daisy Chain

This Part is primarily about how to wire up and install a "1st generation" daisy-chain wired system.

- Wiring
- Connectors
- Power
- Space Envelopes
- Mounting Principles

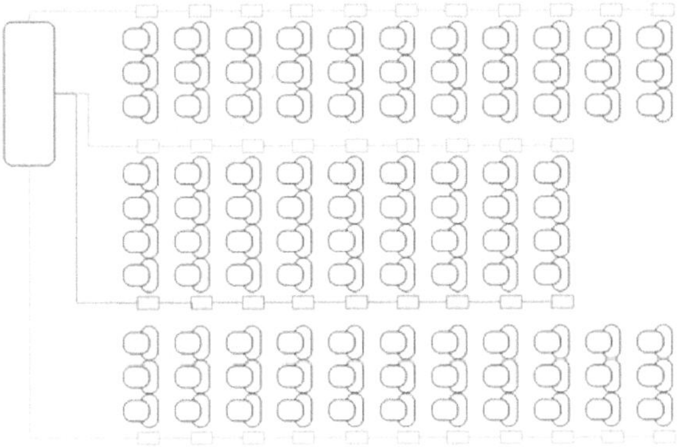

Figure 88 - Daisy-Chain

Part 4B – Cabin Management and Entertainment System – Cabin Distribution System - Star Wiring

This Part is primarily about how to wire up and install a star-wired system.

- Wiring
- Connectors
- Power
- Space Envelopes
- Mounting Principles

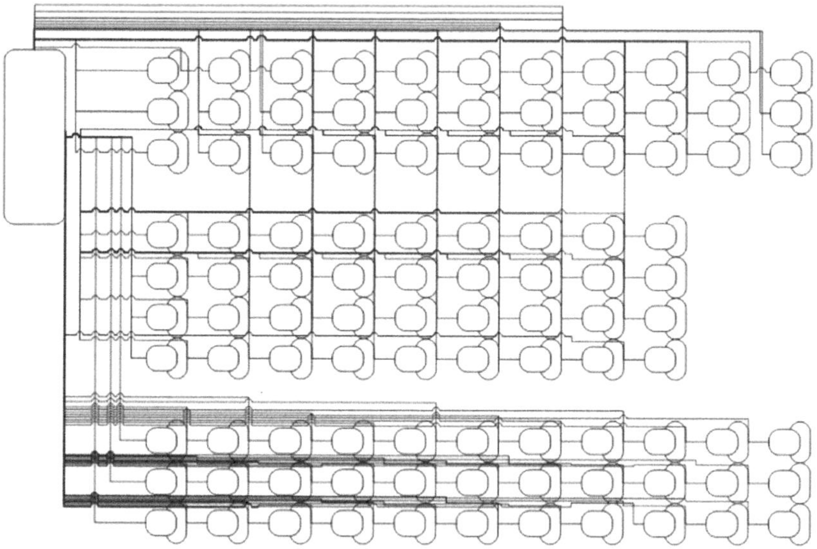

Figure 89 - Star-Wired

Part 4C – Cabin Management and Entertainment System – Cabin Distribution System – (2nd Gen.) – Daisy Chain

This Part is primarily about how to wire up and install a "2nd generation" daisy-chain wired system.

- Wiring
- Connectors

- Power
- Space Envelopes
- Mounting Principles

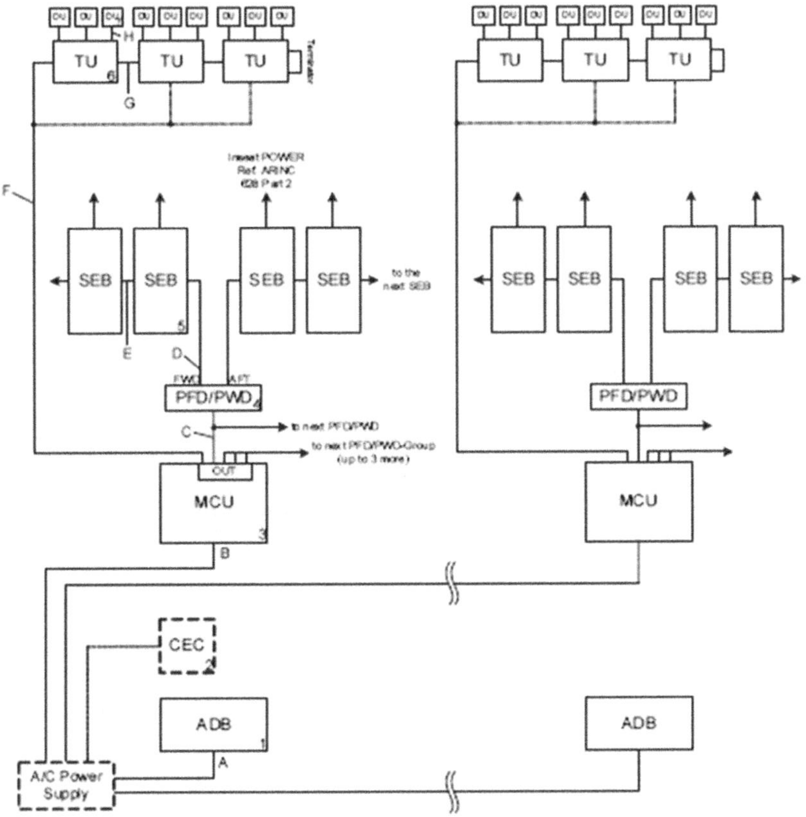

Figure 90 - 2nd Generation Daisy-Chain (Power)

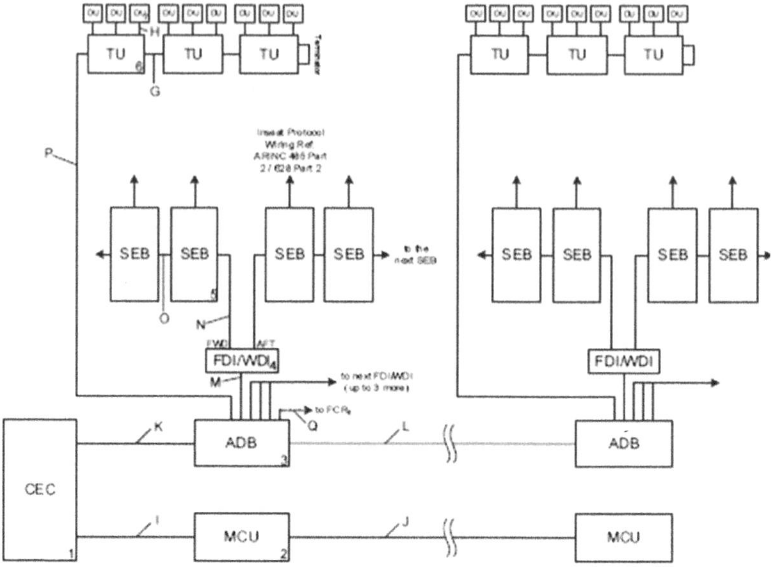

Figure 91 - 2nd generation Daisy-Chain (Data)

Part 5 – Cabin Electrical and Wiring Installation Guidelines

This part is all about how to properly install the wiring and cabling used to interconnect all the various electronic boxes.

- Parts Selection
- Minimum Wire Gauges
- Wire Markings
- Seat-to-Seat
- In-Seat
- General Installation
- Documentation

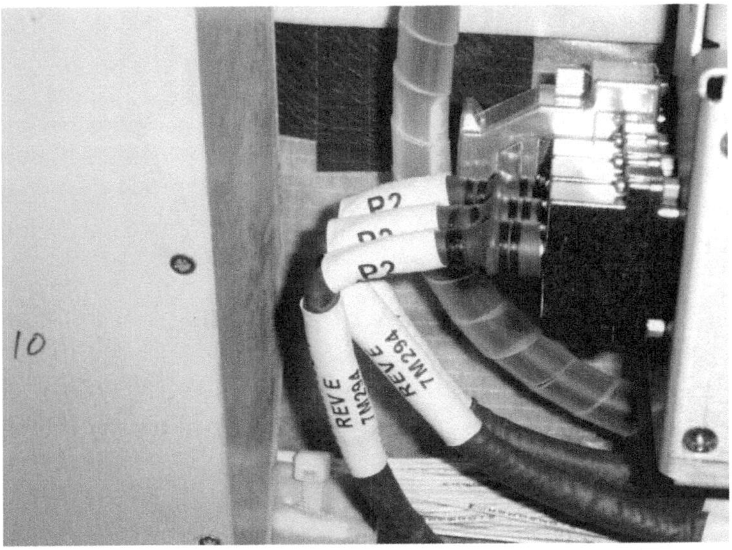

Figure 92 - Cables Bending

Part 6 – Cabin Management and Entertainment System – Fiber Optic Cable Assembly General Specification

This part talks about cabling as well but specifically the proper installation of Fiber Optic cabling.

- Cable Types
- Cable Design
- Identification
- Service Life
- Smoke/Toxicity

Figure 93 - Fiber Optic Cabling

A couple of the key points it address are also different connector types, butt or expanded beam, as well as Single-Mode and Multi-Mode fibers

Part 7 – Cabin Equipment Cooling General Specification: Appendices A, B & C

This Part explains how the cooling aspects of various equipment needs to be addressed.

- Forced Air Cooling
- Area Cooled
- Stand-Alone Cooled
- LRU Cooling Classifications
- Smoke Detection

Figure 94 – This fan is too big.

Basically, how do you get the heat out of the box?

This part is immensely useful when you are looking at determining the Air Conditioning capabilities of the aircraft. Can you keep the cabin cool with the system running?

Part 8 – In-Flight Entertainment (IFE) Equipment Standard Availability Measurement Guidelines

- Availability Calculation Approach
- Definitions of Functional Groups
- Definition of a Failure
- DATA Sources
- BITE

AVAILABILITY is DIFFERENT than RELIABILITY.

Let me say that another way, availability is not reliability.

- 95% <u>reliability</u> means 20 people on a B747 have a non-functioning IFE.

- 95% availability means that 5% of the "functionality" is not available, which may or may not be noticeable by the passengers.

Part 9 – Cabin Information Network (CIN)

This Part is all about connectivity.

- Internet
- Intranet
- Email
- Servers
- Data exchange via Wired and Wireless LAN

Other Industry Specifications of Interest

There are a whole host of other industry guidelines and publications that should be reviewed and understood prior to undertaking an IFE system installation. Some of the more notable ones include:

- ARINC Characteristic 765 – Ethernet Switch Unit
- Project Paper 791 – Ku-Band Satellite System
- ARINC Specification 808 – 3GCN Cabin Distribution System
- ARINC Specification 809 – 3GCN – Seat Distribution
- Project Paper 824 – Onboard Mobile Telephone System
- Project Paper 831 – EMC Recommended Practice
- Project Paper 832 – 4GCN Cabin Distribution System
- Project Paper 833 – Wireless Distribution

Consumer Technology Issues

There have been quite a few companies over the years that have simply tried to take some off-the-shelf consumer products and put them into aircraft. That is not typically a good strategy to try for a great many reasons, a few of which are:

- Certification Issues
- Parts Obsolescence
- Necessity of an FAA Certified repair Station

Let's take each of these in turn for a moment.

Certification

In order to get the part certified for use on an aircraft it needs to undergo many rather stringent environmental tests. As explained in the above section on "Regulatory and Certification Issues", the units go through a lot of abusive testing in order to show they are fit for use on an aircraft.

Most consumer goods simply cannot stand up to the environmental stresses that can occur in an aircraft environment. That coupled with the Electromagnetic Interference (EMI) testing, it's a rather difficult proposition.

Although it's certainly not impossible, you can "repackage" many types of electronics equipment into new housings to mitigate a lot of the issues inherent in consumer goods.

This could then lead to decisions on repackaging the "guts" of a device or taking the entire consumer product, as-is, and wrapping a new metal enclosure around that. Both possibilities have their own unique set of problems.

Parts Obsolescence

In case you haven't noticed, the consumer industry has a very fast paced and volatile product life cycle. Where product lives are counted in months rather than years/decades as is the norm in Aerospace.

So, the odds are very high that if you find a consumer product that can be adapted, modified, repackaged to be certified for aircraft use, by the time you finish your certification efforts that part will probably no longer be in production. So you had better buy a whole bunch of spare units and/or units that you can store and sell to additional customers.

There is no such animal as a consumer product that stays the same for 15-20 years.

Even for products that are very similar in design, "similar" is not the same as "identical". In the aviation world, if a part changes its "Form", "Fit" or "Function", the part number has to change. So if a consumer product has a connector position change, that's a part number change. If a consumer product has an increase in memory/storage that would make its functionality no longer backward compatible, that's a part number change. Heck, if the COLOR of the unit changes, that's a part number change.

ANY change in the form, fit or function means the part number has to roll. And when you change part numbers you have another whole certification effort to go through. Granted, some changes are minor and possibly less testing would be required, but you still need to "document" all of the differences and similarities to currently certified products if you don't want to go through and re-test everything. Then you have to get that documentation approved by the FAA.

FAA Repair Station

And, since most companies that are trying to use consumer goods typically don't have these parts on their Part 145 repair certificate, when there is any failure within the unit, you must scrap the entire thing.

You can't send the part back to a consumer electronics company and have them repair it and then put it back into an aircraft. They are not a part 145 certified repair station. There is no oversight that the processes they have in place are repeatable and can guarantee airworthiness.

Take EMI for example, most electronic goods have shields internal to the case to restrict or reduce the unwanted emission of RF energy. During the repair of an item, the consumer electronics technician may forget to replace them or they may lose them or it may be a hassle to put these shields back on. So the unit comes back from repair working, but it may no longer pass the strict EMI emission controls that are in place for aircraft. So this unit could be way out of tolerance, spewing disruptive RF signals everywhere, and the supplier would never know.

There have been a lot of pieces of consumer equipment being re-packaged and called airworthy over the years. If you are thinking about buying one of these products, be cautious and ensure that the equipment is tested to the appropriate DO-160 categories. And that the company has repair policies and procedures in place that will satisfy all of the appropriate airworthiness requirements.

Having said that, don't get too crazy with some of the requirements. An LCD monitor is not going to function at minus 50°C (-50°C) as the "L" part of LCD is "liquid" and the display would probably be frozen at that temperature.

BUSINESS and CORPORATE JETS

Air Transport versus Business Jets

Okay, now that we have covered a fair bit of territory on the Air Transport side of things, let's chat a bit about smaller jet aircraft. These are the companies most people think about when they think of Business Jets: Gulfstream, Bombardier, Dassault, Embraer, Hawker Beechcraft, and most recently Honda Jet. There are certainly others of course, but these are really the ones that I hear the most about.

Figure 95 – Some Business Jets

Business Jet

Business and Corporate Jets will collectively be called "Business Jets" from here on out.

Business Jets have some unique characteristics that immediately set them apart from their Air Transport cousins.

The "category" of business jets typically restricts them to a maximum of 18 passenger seats. Although very few actually fly with this number of seat places installed.

The Business Jet category is generally broken down into five (5) sub-categories, depending on the size of the aircraft.

- Very Light Jets
  - 3-6 Passengers
  - Typical Max Range: 1,000 – 1,500nm
- Light Jets
  - 6-8 Passengers
  - Typical Max Range: 1,250 – 2,500nm
- Mid-Size Jets
  - 8-10 Passengers
  - Typical Max Range: 3,000 – 4,000nm
- Super-Mid-Size Jets
  - 8-14 Passengers
  - Typical Max Range: 4,000 – 5,000nm
- Heavy/Large Cabin Jets
  - 12-18 Passengers
  - Typical Max Range: 6,000 – 7,000nm

Now, before I get too much further, I'd like to mention turbo-prop aircraft. There are some amazing aircraft in this area like the King Air 350, Pilatus PC-12 and Cessna Caravan family just to name a few. These, and others like them, are all really great and amazing aircraft. Unfortunately, when it comes to IFE, entertainment isn't typically a high priority. The focus is typically on the cockpit avionics and the bits that actually make the aircraft flyable. Entertainment systems are too frequently just an afterthought.

That is not to say that none of them have IFE, or as it's called in the Biz Jet world, a Cabin Management System (CMS). Many customers opt to install CMS into these aircraft but the vast "majority" do not. Most have simple reading light controls for the passengers and that's about it.

Newer versions of these fine aircraft, like the 350i, have an option for High-Definition CMS systems that include 15" LCD monitors.

So, perhaps as time moves on and the buyers of this type of aircraft become as discerning about the passenger cabin as they are about the cockpit avionics, more turbo-prop aircraft will install CMS systems. I for one am looking forward to that day.

And now back to our regularly schedule discussion...

Despite the name, Business Jets for the most part are very personal. The interiors are designed and configured individually for each buyer. There is no one-size-fits-all when it comes to these aircraft.

Every interior is unique. For example, two different aircraft may share a common seat type, as far as structure goes, but the look may differ entirely due to choices of cushions, color, fabric, leather, trim, gold embellishments, piping style, etc.

Also the sidewall panels, ceiling, galley inserts and locations are typically all open to some form of customization. Additionally, there is nearly infinite variety and customization of the CMS system, which in this type of aircraft comprises far more than just entertainment.

With the CMS in a business jet the passenger can monitor and control an enormous array of devices and functions from cabin temperature, to cabin lighting, to the audio and video programs playing in various zones of the aircraft to heated seats, heated floors, massage chairs and unique custom and one-of-a-kind devices and systems.

These aircraft have a very elaborate equipment feature set indeed. While some may be content with being on the "Leading Edge" of technology, frequently these buyers want to be on the "Bleeding Edge" of technology.

Very, Very Different from Air Transport

With Air Transport, the passengers are passengers and they are carried around by a transportation company.

In the business jet world the passengers are the owners of the aircraft, or at least they think they are. And in reality, in almost every case, no one person actually owns the business jet they are flying in.

These aircraft are owned by corporations. Either big corps or little corps, it doesn't matter, they are owned by corps. There are huge huge tax benefits for people not owning the aircraft themselves, as an individual. Some people even go so far as to set up corporations that exist solely to operate their aircraft.

But, and this is a big but, many CEO's and Presidents of these companies tend to think of the jets as their own private vehicles. More so even than say a company car. The corporate aircraft tends to become an extension of themselves. There is a tremendous amount of prestige that goes along with operating a business jet.

That is absolutely NOT to say that these are frivolous and merely vanity items. These aircraft absolutely benefit the bottom line of most companies when they are used to carry employees to trouble-spots quickly to solve problems for their clients.

Additionally, some operators of these aircraft allow conversions to medical transport use by taking out seats and putting in hospital style beds when the aircraft are not in use by the owners. An incredible humanitarian benefit to be sure.

In Air Transport aircraft, the airline typically provides the entertainment programming and content whereas in

Business Jets the passengers bring their own. And here I don't mean bring their own as in they have a portable A/V player. Here I mean they bring on-board their own DVD or Blu-ray movies, USB sticks with A/V content, etc., and play the movies or listen to MP3s over the Business Jet's A/V system. Entertainment content "variety" here is probably one of the primary benefits of the CMS. Most of these aircraft have a small bank of several Blu-ray players so that many movies can be played over the system at the same time.

And while content variety is very important, the aesthetics of how the system looks is critically so. In some cases how the equipment looks can be even more important than how well it functions.

Figure 96 - Business Jet Interior

Another thing that also sets this market apart is the frequency of last minute changes. One morning you'll think you're finishing up the interior build, but that afternoon the principal, or their representative walks through the cabin, and the next day you're trying to figure out how to remove a table with four chairs and replace them with a divan.

Last minute changes like this are unfortunately far too common.

As I said before, every aircraft is unique. Except for those that aren't. And by that I mean some companies have "fleets" of business jets, like NetJets and Flexjet/flightOptions, and they tend to want them to have as close to the same amenities as possible. This is so that the fractional owners, charter members, and pass card participants can have a similar flying experience no matter which aircraft, within the same category, they are flying on.

Figure 97 - Business Jet Interior

Figure 98 - Pilatus PC12 (Turbo-Prop)

Okay, so the above isn't a business "jet", but it is a business turbo-prop with a nice embedded CMS system installed. So I thought I'd include it in the picture gallery. This is certainly not typical of this aircraft type but it is a nice CMS implementation.

Next are a few Cabin Layouts. There isn't really such a thing as a standard layout, or a typical layout as every aircraft is different. But these should give you an idea about the types of features in the interior that can be expected.

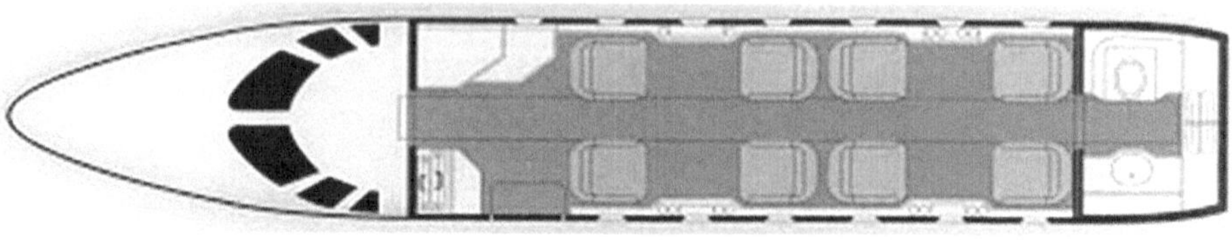

Figure 99 - Business Jet LOPA

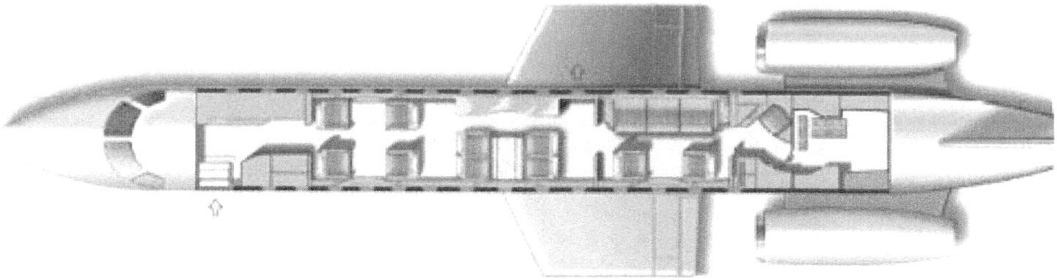

Figure 100 - Business Jet LOPA

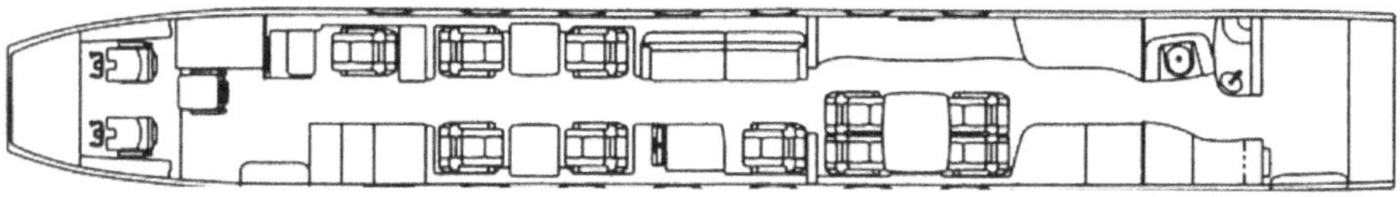

Figure 101 - Business Jet LOPA

Major Emphasis in this Market

There are four (4) primary areas that people and companies focus on in this market.

- System Weight
- System Configurability
- System Installation
- System Cost

All of these items have varying degrees of importance depending on the customer.

System Weight

Weight is a key factor in pretty much everything related to aircraft. The heavier something is the more power and more fuel it takes to lift and carry it from one place to another. Fuel burn affects the cost to operate the aircraft, as more fuel consumed equals higher operating costs.

Additionally, weight directly affects the aircraft performance characteristics. I know of one aircraft manufacturer that replaced all of the circuit breakers in the cockpit when they found they could save 9 grams (0.32 ounces) per breaker.

But keep in mind that the performance characteristics of an aircraft are calculated when empty. So those extra 9 grams per breaker translated directly into increased range and/or increased payload on the aircraft spec sheet.

System Configurability

As previously mentioned, changes to the aircraft configuration happen all the time. This primarily applies to completions or during the aircraft refurbishment process.

The systems need to be able to accommodate last minute changes to source equipment, monitors, location changes, etc.

If someone has a table and 4 chairs going in, and then changes their mind and want a 3 piece divan in that spot instead, the system needs to be able to handle that with minimal changes. Some systems on the market still require massive amounts of cabling to be ripped out and re-laid but thankfully they are in the minority.

System Installation

The equipment that makes up the system needs to be designed for quick and easy installation; this also includes the wiring plant.

Typically, a remove/replace time of less than 5 minutes is levied on the suppliers for any one of their boxes. However, that time limit does not include getting access to the box, just the remove/replace time of the box itself.

I have seen some boxes fielded that had 20+ circular connectors fitted. It took longer than 5 minutes just to get all the connectors off. The connectors were spaced very close together and that made it difficult for the mechanic to put their fingers around any one connector. Not a very good box design at all.

Also, many of the units are exposed to the passenger, (e.g. PCU, Monitor) and these need decorative bezels. One issue that comes up time and again is the installation of PCU's and monitors, with the decorative bezels installed, while testing the aircraft wiring and system functions. This is typically done prior to the rest of the aircraft cabinetry being finished. So, with a gold plated PCU installed in a side-ledge, and mechanics moving cabinetry around them, these plated surfaces routinely get scratched or dented. The

PCU/Monitor must then be removed and sent back to the manufacturer for re-plating. Unfortunately, for the initial fit all plated objects are done at once so as to ensure a common color. When only one or two items are removed for re-plating, it's difficult to get an "exact" match.

System Cost

Believe it or not, cost is still an issue for these systems. More of an issue for some people than for others, but it's still an issue.

Retrofit Versus OEM Installation

There are similar issues here to what I mentioned in Air transport but they are not nearly as divergent.

Although here to the OEM wants to make "standard" aircraft but since every aircraft customer is unique, that is a difficult proposition. So what many of them now do is to provide a catalog of many different cabin configurations to choose from and then the work the rest of the installation from that baseline.

This still gives the buyer a very large selection while keeping the overall engineering and finishing costs reasonable. That is not to say that if someone wants to pay for something totally unique that the Business Jet OEMs would turn them away, far from it, but they do want to gain the benefit of standardization whenever possible.

This is true for both the cabin interior layout and the IFEC systems as well. All OEM's have their "preferred" system and vendor. But I believe all of them are still open to working with any of the suppliers given the right incentive to do so.

Also, several of the OEMs actually break up the delivery cycle into two parts, one is the "Production" build of the basic aircraft, and the other is the "Completion" of the aircraft.

The completion phase includes all of the unique interior elements, galleys, IFEC systems, external antennas, and etc., etc., etc. Other OEMs do this like the Air Transport OEMs do; they handle all of the completion activities on the production line.

I believe that breaking up the production phase and the completion phase into two separate processes is probably better overall in the long run. It lets the OEM "build" every aircraft the same and then move it to another location where they have specialists that can do all of the custom modifications.

IFE/CMS System Functionality

In your reading through the section on Air Transport aircraft, with all the audio, video and communication options, you may be thinking that Business Jets is more of the same. In that, you would only partially be right. Yes, there is Audio, Video and Communications, but there is so much more that it's mind-boggling.

At a bare minimum, you typically have the same "types" of functionality as in Air Transport:

- Audio and Video Distribution
  - Typically Broadcast only and not AVOD
- In-Seat Reading Light Control

And then we get into some things that are not passenger controllable in a standard commercial aircraft:

- Audio and Video "Transport" Controls
  - Start, Stop, Play and Rewind for the Cabin Monitors AND in the In-Seat Monitors.
- Cabin Environmental Controls
  - Cabin and Seat Heating/Cooling
  - Floor Heating
- Cabin Lighting
- Table Lights
- Mood Lighting
- Motorized Window Shades

And speaking of motorized gizmos:

- How about a motorized Vodka dispenser?
  - For one program, the principle pushed a button and a bottle of Vodka came up through an opening hole in a freezer top. He pushed the button again and it went back into the freezer and the lid closed behind it.

- ✈ And one client that wanted a toilet flush button added to the Cabin Remote Control.
    - o Apparently quite the practical joker...

Now, one thing of interest that I'm not sure you caught was when I said that the Video was typically "Broadcast Only" as opposed to "Audio/Video on Demand" (AVOD). Why do you think a Business Jet Owner wouldn't like a hard drive full of movies to watch? Go ahead and think about that a minute, I'll wait...........

Ready? Because on Business Jets, VIP and VVIP Aircraft they have installed many DVD/Blu-Ray players and bring aboard the movies they want for "that flight" and don't have the time and expense of buying servers pre-loaded with movies. Or have to wait for a contracted service to load the next month's movies on-board.

If they have an AVOD server on-board with 100 movies, it's still the SAME 100 movies they had on their flight yesterday, and the day before that, and the day before that... "Variety" is critically important as they want to watch whatever strikes their fancy at the moment. Sending someone off to the buy a video and dropping the disk into the player has proven to be infinitely more convenient that using an AVOD server.

Now, some of you may be familiar with a little regulation in the United States called the "Digital Millennium Copyright Act".

For the intent and purpose of this book, I'm not covering that regulation at all. There is a plethora of information available should you wish to dig deeper.

Related System Functionality

Similar systems to those found in Air Transport aircraft include:

- Air to Ground & SATCOM Communications
  - High-Speed Internet Access
  - E-Mail
  - Voice Communications
- Moving Map Functionality
  - Typically much higher resolution than in Air Transport
  - Many systems allow the passenger to have a high degree of control over what's being displayed.
  - Many customers have a "Mecca Pointer" added to the display so that always know in which direction Mecca is.
  - Many other customers add custom pointers to the corporate HQ, or their vacation home, or whatever strikes their fancy.
- Security Systems and Wing/Belly Cameras
  - These systems provide video entertainment while in-flight and provide security monitoring and alerting while on the ground.

Figure 102 – Securaplane Security System

Now on to one of my favorite topics, Video Games.

They are close to, but not the same as, Air Transport.

- Business Jets
  - Game consoles are Carry-On Equipment (PEDs)
  - Always have the latest technology available
  - XBOX 360 – XBOX ONE – XBOX 3,000
  - Playstation 3 – Playstation 4 – Playstation 50
  - Plugs directly into one of the aircraft video monitors and audio system or into the IFE distribution system
  - No FAA certification or substantiation required. Just carry it on, plug it in, and play.
  - Always have the latest and greatest technology available.

They just carry-on the game console, plug it in and away they go. When the next big console gets released they just buy it, bring it on-board, and plug it in.

Business Jet System Design

The players in this market are a lot different, more numerous and varied than in Air Transport, mostly because there are so many more aircraft manufacturers in this space. And of course the custom nature of this market sector can support a lot of players.

The architectures of the systems are also quite different. Figure 103 shows one of the many systems I helped to design for this market sector. It uses several innovations that reduced the amount of wiring, the system weight, and the installation and maintenance costs.

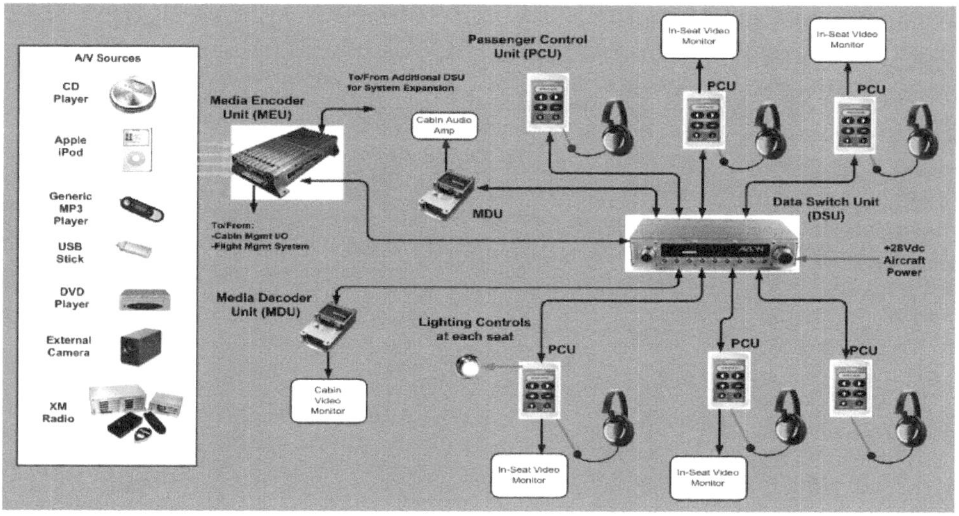

Figure 103 – Business Jet IFE

Many other system topologies are in use from star-wired HDMI to daisy-chain networks. With the HDMI star-wiring typically using stock or sometimes modified commercial-off-the-shelf (COTS) consumer hardware. If you refer back to Figure 2 and Figure 3 you'll get the idea. The big difference here of course is that these aircraft are so much smaller than the air transport planes that the weight impact isn't as great. Although there is an impact and I said before, weight is a critical factor in these aircraft.

Also, please recall that these systems control many many more things than the systems installed in Air Transport class aircraft. It's not just reading and call lights and IFE. They call these Cabin Management Systems (CMS) for a reason.

And on that note, next is a very small sample of the types of controls that the passengers have access to in a Business Jet. Again, this is by no means a comprehensive list; it is here to give you a flavor of the scope of what passengers have control over.

Many aircraft will also include a whole slew of custom controls like coffee warmers, seat warmers, seat massagers, and far more lighting types than covered in the following tables.

Some Typical Business Jet PCU Controls

LIGHTING	
	In-Seat Reading Light (Local)
	Table Light (Local)
	Cabin Lighting (Forward)
	Cabin Lighting (Mid)
	Cabin Lighting (Aft)
	Cabin Lighting All
	Mood Lighting (Forward)
	Mood Lighting (Mid)
	Mood Lighting (Aft)
	Mood Lighting All
	Entry Light
	Galley Lighting
AUDIO	
	Display of Current Headset Audio Source Selected
	Headset Audio Source Select
	Headset Audio Volume
	Headset Audio (Bass/Treble)
	Display of Current Cabin Audio Source Selected (Fwd)
	Cabin Audio Source Select (Forward)
	Cabin Audio Volume (Forward)
	Display of Current Cabin Audio Source Selected (Aft)
	Cabin Audio Source Select (Aft)
	Cabin Audio Volume (Aft)
	Transport Controls: Play
	Transport Controls: Stop
	Transport Controls: Fast Forward
	Transport Controls: Rewind
	Transport Controls: Skip Song/Track
	Transport Controls: Pause

VIDEO	
	Display of Current Video Source Selected (Local)
	Video Source Select (Local)
	Video Brightness (Local)
	Display of Current Video Source Selected (Fwd Cabin)
	Video Source Select (Fwd Cabin Monitor)
	Display of Current Video Source Selected (Mid Cabin)
	Video Source Select (Mid Cabin Monitor)
	Display of Current Video Source Selected (Aft Cabin)
	Video Source Select (Aft Cabin Monitor)
	DVD Menu Control: Menu Select/Navigate and Select
	Transport Controls: Play
	Transport Controls: Stop
	Transport Controls: Fast Forward
	Transport Controls: Rewind
	Transport Controls: Pause
	Transport Controls: Next Chapter Mark
	Transport Controls: Previous Chapter Mark
ENVIRONMENTALS	
	Cabin Temperature (Display) (Forward)
	Cabin Temperature (Up/Down)(Set Temp) (Forward)
	Cabin Temperature (Display) (Aft)
	Cabin Temperature (Up/Down)(Set Temp) (Aft)
MISCELLANEOUS	
	Window Shade (Up/Down)(Start/Stop) (Local)
	Window Shade (Up/Down) (All)
	Moving Map Function/Display Program Mode
	Moving Map mouse controls (Navigate and Select)
	Flight Attendant Call Button/Chime
	Toilet Flush

Early Business Jet PCU's used dedicated switches for each function and if the principal added anything, the entire switch needed to be redesigned and the old one thrown away. A VERY costly work flow.

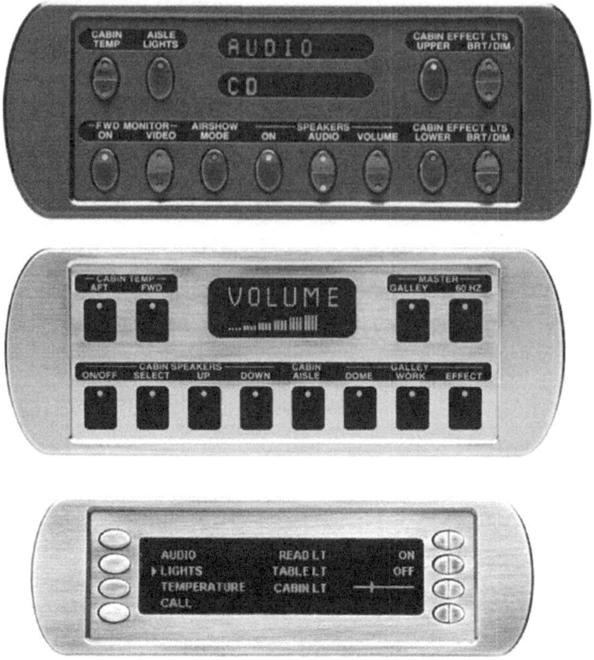

Figure 104 – Older PCU/Switch Panels

Figure 105 – Very Custom, Button Style, Switch Panel

Then, as technology improved, PCU's that used touch-screens became more the norm. Now you can add and remove functionality by just changing a few lines of code.

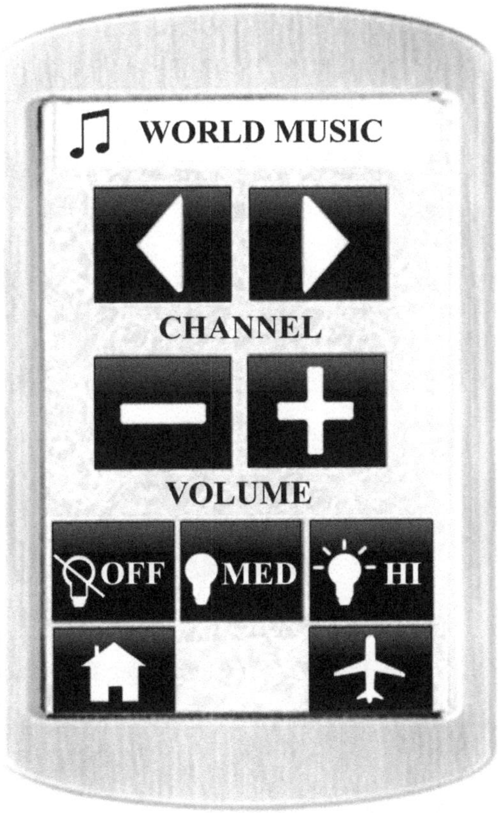

Figure 106 - 3.7" Touch-Screen PCU

FA Panel and Handheld Remotes

Built-In Flight Attendant panels and GUI's on Tablets that sync and allow control of all aspects of the cabin are readily available in this sector. They come in every type and style imaginable. Below are a few examples.

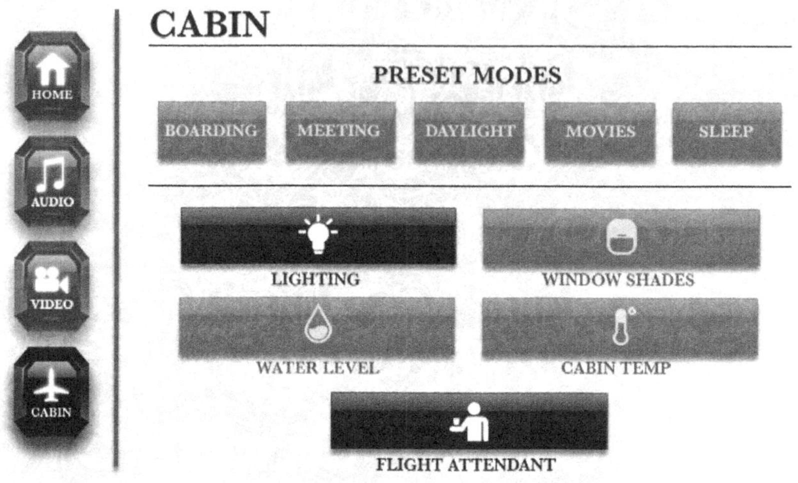

Figure 107 - FA Panel Graphical User Interface (GUI)

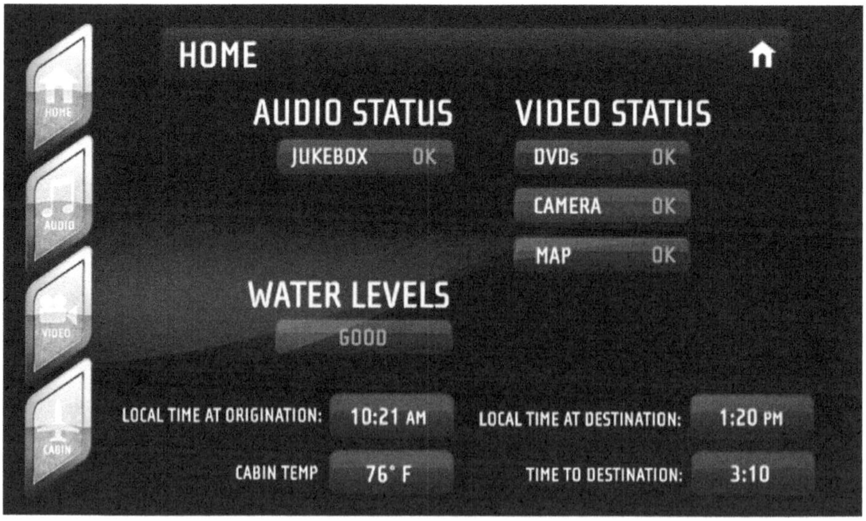

Figure 108 - FA Panel Graphical User Interface (GUI)

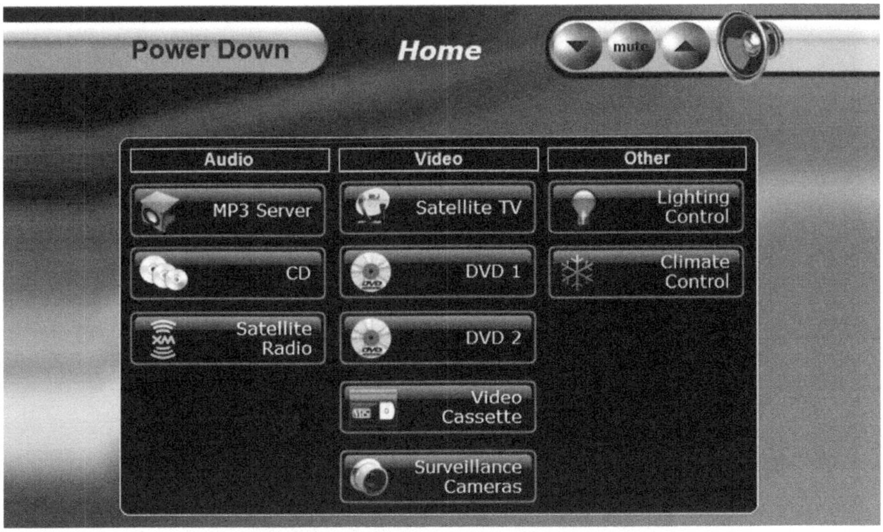

Figure 109 - FA Panel Graphical User Interface (GUI)

LCD Cabin Monitors

There are a huge variety of shapes and sizes and mounting methods for monitors in these cabins.

Figure 110 - At Seat 10.4", 8.4", 7" Wide, 6.3" LCD

Figure 111 – 15" to 24" LCDs in various aspect ratios

Figure 112 - Plug-In Monitor Bases

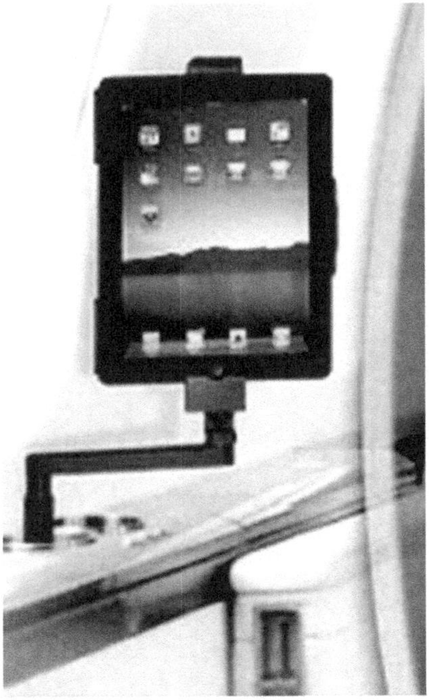

Figure 113 - iPad 2 Arm Mount

There are even mounts made that use an existing Rosen plug-in port to hold up an iPad instead. No control lines to it though, just a support piece so you don't have to hold it. Although some mounts are available that can convert the aircraft power to charge the iPad.

SiriusXM Radio

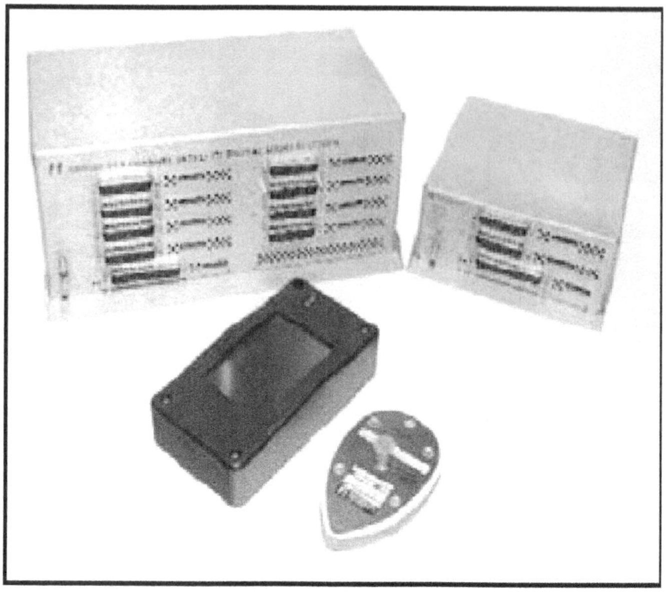

Figure 114 - XM Radio

Satellite Radio is VERY popular in this market. Especially the XM Weather which can give the cabin a free satellite audio channel, but with the budget for it coming from the Avionics Budget as opposed to the Cabin Budget.

Source Equipment

Traditional Source Equipment

I mentioned before that instead of AVOD servers these aircraft typically use DVD/Blu-Ray players. Believe it or not there are still many of these aircraft fitted with VHS tape VCR's.

Additionally, since the FAA hasn't historically been as strict in this market, we also find many repackaged consumer goods, like car stereos and other A/V devices.

Figure 115 - Repackaged Automotive Equipment

Figure 116 - Repackaged VCR

Business Jet Newer Source Equipment

Here is a bit more about the more recent Source Equipment types:

- DVD/Blu-ray players based on PC Computing standards
    - Allows for true digital end-to-end transmission
- Game consoles like the PS4 being used as a Blu-ray video player
- MP3 player support
    - Direct Digital Interfaces
    - Compact Flash and SD card readers
    - USB Stick readers
- Allows users of touch-screen display PCUs to select music via meta-data
    - Artist, album, genre, etc.

Business Jet Cabin Audio

Another place that Business and VIP/VVIP aircraft differ from Air Transport aircraft is the Cabin Audio System.

In these aircraft it is actually quite rare to have the executive wearing headsets. Usually their audio is supplied by cabin speakers, usually using surround-sound processors. Keep in mind that these aircraft are typically divided up into smaller zones (rooms) by walls and other in-cabin monuments. They also tend to have very large bulkhead mounted video monitors so they can watch movies as they would in their home. That is, not wearing a headset.

Now, this is NOT the less than optimal, to put it mildly, audio that you find from the Public Address (PA) system on commercial airlines. The garbled voice of the pilot telling you about some non-descript geographical location hidden somewhere under the cloud cover. Or was it the latest football score? I couldn't tell...

These executive jets offer sound quality that is far better than most home sound systems. And since the executive jet interiors are orders-of-magnitude quieter than their commercial airline counterparts, you can actually hear the pristine audio that these systems deliver.

The companies that operate in this space take the aircraft LOPA (Layout of Passenger Area) and help to design speaker locations, types of speakers, and sub-woofer quantities and locations. Taking into account not just the general room layout but also the effect of fabric on the walls and even large flat spots that may impact the audio reflections.

Alto Aviation is one company in this market and they configure the aircraft interior sound system by setting up microphones around the cabin and then monitoring the sound at each location. They then tweak their system by watching what's happening on the electronic scopes. This is an interesting and very visual process to adjusting the sound.

Companies in this area also design and built their own audio system components which include some very interesting speaker and amplifier designs.

You would be amazed at the sound quality that can come out of some of the small speakers these companies come up with. How they can move that much air to create such amazing sound is really a mystery.

Business Jet Connectivity

Here again is where there is some bit of compatibility and a whole lot of difference with Air Transport.

- SATCOM
  - Most of these aircraft will have SATCOM systems installed for off-board communications
- Terrestrial Broadband
  - Domestic US aircraft may install a Terrestrial Broadband system instead of, or in addition to, the SATCOM system
- Satellite TV
  - Many aircraft will have small SAT TV dishes installed in the tail (at the top of the vertical stabilizer).
- Phone Systems
  - These will normally be part of a dedicated phone system installed in the aircraft that combines some number of wired and wireless handsets.
  - Outside the US some aircraft are equipped with systems that allow passengers to use their own, personal, phones while airborne. Same as described in the Air Transport section.
- Facsimile (FAX) Machines
  - Believe it or not, there are still many business aircraft that require this functionality.

Business Jet "Built-In" Phone Systems

I mentioned this briefly above but wanted to expand on it a bit.

These systems are dedicated phone systems with PBX switches and handsets and all the goodness that makes an office phone system work on the ground, and they bring that functionality to the air.

As you may imagine, there are several companies in this space. Another disclaimer here about this not being a comprehensive list, etc.

- GoGo
- ICG
- TrueNorth Avionics

These companies have dedicated wired and wireless phone options and can typically use whatever off-aircraft link is available. Typically these include Iridium, Terrestrial Broadband and others.

Figure 117 – TrueNorth Phone

The picture shown in Figure 117 is from the TrueNorth customizer on their web site. Pretty nifty tool to get a feel for how different customized backs may look in various interior styles.

Business Jet Phone Systems for GSM Cell Phones

Some of the same players are in this market and some new ones as well. For example, the graphic in Figure 118 is from OnAir and explains a bit about the path your phone call goes through to get to its final destination.

Figure 118 – OnAir GSM Phone Call Path

Pretty much these systems can work with any off-aircraft communications link. The key piece here, for all of these systems, is the on-board pico cell. Basically a teeny tiny cell tower in the aircraft that talks to the cell phones on-board. It then converts those phone signals to work with whatever off-aircraft communications link is installed.

This serves several purposes, one being that it automatically forces your phone to switch to its lowest transmit power setting. This greatly reduces any chance that your phone could connect to a cell tower on the ground. It also helps improve your battery life as it's transmitting at such a low power.

Second, although not an issue here in Business Jets, but in Air Transport aircraft remember it lets the airline pocket a piece of the roaming charge. However, here you pay the off-aircraft communications provider directly for whatever air-time you want to buy. And it's still not cheap...

Business Jet and VIP/VVIP Security Systems

These systems form a dual purpose, one is as video feeds for the IFE while in-flight and two is for security while on the ground.

First let's talk about the IFE functionality. These systems typically have a multitude of cameras installed looking at various portions of the aircraft exterior.

Typical cameras and their locations include:

- Quad-Belly Camera
    - This housing includes four cameras each pointing in a different direction with overlapping fields of view.
        - Front Target: Nose Gear
        - Right Target: Right Wing and Engines
        - Back Target: Main Landing Gear
        - Left Target: Left Wing and Engines
    - Depending on the size of the aircraft, one assembly may be sufficient but on larger aircraft a second unit may be needed behind the main landing gear to ensure coverage of the gear and the engines.

- Tail Camera
  - Typically installed at the top of the vertical stabilizer looking forward. This gives you a very good view of a significant portion around the aircraft.
- Zoom Camera
  - These are typically installed in the belly and point straight down. They also have variable zoom capability.
- Tilt/Zoom Camera
  - These are also typically installed in the belly but they let a passenger steer and zoom the camera at will.
- Flight Deck Camera
  - There are several types of cameras available.
    - One type goes on the bulkhead behind the pilots so that you can see the back of their heads, some of the instruments and out the windshield.
    - Other types fit right to the windshield for an unobstructed view out the front.

For security purposes on the ground, the Quad-Belly Cameras are usually fitted with Infra-Red motion detectors (Active or Passive) and the video feeds themselves can also be used to detect motion. And while normally these systems are simply used as HD color cameras, they do have options for night-vision cameras which are more typical of military security situations.

The rest of how the security systems work, how they communicate in real-time with security personnel and all the different types of wired and wireless sensors is beyond the scope of this book. But, I can say that for aircraft that have these systems installed, DO NOT try and get close to them if you are not authorized to do so.

Although in most cases I've heard about, the intent of people messing with the aircraft hasn't been sabotage or malicious intent, but drug smuggling...

In any case, if you have a business jet, and fly out of the country, I highly recommend installing one of these systems.

VIP & VVIP AIRCRAFT

VIP & VVIP Aircraft Systems

I've mentioned VIP and VVIP aircraft before, but what is a VIP/VVIP aircraft?

For the purposes of this book, I'm lumping them together and defining them with the following characteristics:

- VIP = Converted Air Transport class aircraft
  - B737, B757, A320/A321, Various Regional Jets, etc.
  - Typically "Single-Aisle/Narrow-Body"
- VVIP = Converted Air Transport class aircraft
  - B787, B747, A380, A340, etc.
  - Typically "Twin-Aisle/Wide-Body"
- Typically Heads of State or very wealthy individuals
- Very Elaborate Interiors
- Aesthetics are usually more important than the functionality
- Many Have Mixed Cabin Configurations
  - VIP/VVIP + Air Transport
- Mostly the same feature sets as in a Business Jet but much larger in scope
  - More destinations: seats, monitors, lighting
  - Same diverse A/V and Cabin Control function options
- Some seats in the back of the aircraft may be in an Air Transport configuration
  - Typically First or Business Class seating with typical Air Transport IFE
    - For Press Corps or Assistants/Staff

The following is an example of a possible B747 configuration. Notice that the front of this aircraft has a

VVIP configuration while the back of the aircraft has a more traditional Air Transport seating arrangement.

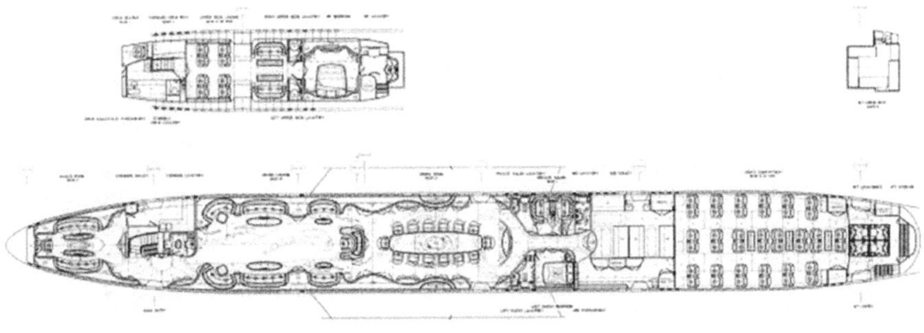

Figure 119 – B747 VVIP LOPA

And here is a graphic of the Embraer Lineage 1000 showing one of its interior configurations.

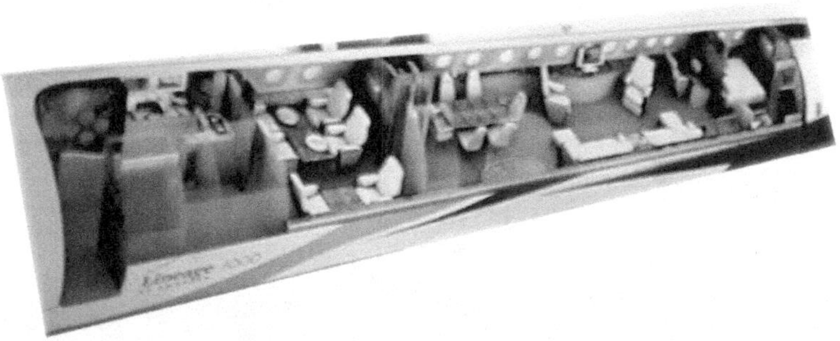

Figure 120 – Embraer Lineage 1000

And now for a few pictures and illustrations of some VVIP interiors.

Figure 121 - VVIP Interior

Figure 122 - VVIP Interior

Figure 123 - VVIP Lavatory

Figure 124 - B787 VVIP Interior

As far as IFE and CMS goes, it's mostly just more of what we saw in the Business Jet section, only more of it. A LOT more of it...

These aircraft also have huge video monitors, sometimes 65" or larger. In the big conference rooms you will usually find one at either end of the cabin and sometimes one along the sidewall covering up some of the windows.

You will also find large video monitors popping up through cabinetry on motorized lifts. Some of the master suites are popular places for this functionality.

Installed Private Branch Exchange (PBX) type phone systems are also the norm here so that passengers can call from one room to another. But this shouldn't be confused with the Flight Attendant call system that is built into the CMS.

Same with the security systems I mentioned earlier. These aircraft have a lot more access panels on the outside that you may not even be aware of. Better safe than sorry...

And there you have it, VVIP aircraft have a lot more gold trim, mother-of-pearl, and precious and semi-precious accents throughout. They also typically combine the CMS of a Business Jet with the IFE of an Air Transport aircraft in different parts of the cabin and generally just have more of everything.

I once worked on an aircraft like this that had maybe 47 different lights just in the master lavatory. There were bright lights, mood lights, night lights, day lights, under-sink lights, mirror lights, vanity lights, trim lights and etc., etc., etc. And let's not forget the under-floor heaters as you stepped out of the shower... Or the various lights "inside" the shower for that matter... All in all some amazingly beautiful aircraft conversions.

And lastly, in the VIP/VVIP space, the Aircraft OEMs don't typically provide any of the interior conversion activities on their production line. The end users typically contract with special design and modification houses like Greenpoint Technologies for the interior design, engineering, and physical installation of the modifications. When it comes to VIP/VVIP conversions, these 3rd parties are really where the expertise resides, not with the Aircraft OEMs.

In some cases a single modification house can handle all of the activities necessary, in other cases the customer may contract with one company for the design, another for the engineering and a third for the physical installation. Or the customer may contract with a 3rd party to manage all of these activities and companies for them. It just depends on the needs of the customer and the capabilities of the people bidding on the project.

And even things like Star Trek™ style sliding doors that are controlled by motion sensors and are driven by pistons can be found in these aircraft, and they to can be found interfacing to the CMS. In fact, almost all interior doors have sensors on them to ensure that they are either open or closed for Taxi, Take-off and Landing (TTL) as per the relevant FAA regulations.

It's really just amazing how much goes into one of these aircraft conversions and the thousands of interfaces to the CMS/IFE that you would have never expected to be there.

THANK YOU

Thank You

And most importantly, thank you for taking some of your valuable time and sharing it with me. I know you can spend your time in a variety of ways and I'm truly honored that you chose to spend it with me.

And with that, I'll be signing off for now.

Oh wait, before I sign off, just a quick note for you to keep an eye out for my next book which will be on the FUTURE of In-Flight Entertainment and Connectivity. That's where I discuss some innovations that will truly rock the IFEC industry.

Thanks again!

The author may be contacted via www.BriskiConsulting.com at IFEbooks@BriskiConsulting.com or via the publisher at www.SpiralJourney.com.

ACRONYMS

A/D	Analog/Digital
AC	Alternating Current
ACARS	Aircraft Communications Addressing and Reporting System
ADPCM	Adaptive Delta Pulse Code Modulation
APEX	Airline Passenger Experience Association
APU	Auxiliary Power Unit
ATG	Air-to-Ground
AVOD	Audio/Video On-Demand
BGM	Boarding Music Machine
BIT	Built-In-Test
BITE	Built-In-Test Equipment
BTS	Base Transceiver Station
CBB	Connexion by Boeing
CD	Compact Disk
CDMA	Code Division Multiple Access
CFR	Code of Federal Regulations
CIN	Cabin Information Network
CONUS	Continental United States
CoS	Class of Service
COTS	Commercial-Off-The-Shelf
CRT	Cathode Ray Tube
CSMA/CD	Carrier Sense Multiple Access with Collision Detection
CMS	Cabin Management System
DC	Direct Current
DIN	Deutsches Institut für Normung
DVD	Digital Video Disk
DVR	Digital Video Recorder
EE-Bay	Electronics Equipment Bay
ELA	Electrical Loads Analysis
EMI	Electromagnetic Interference
ESD	Electrostatic Discharge
FA	Flight Attendant
FAA	Federal Aviation Administration
FAR	Federal Airworthiness Regulations

ACRONYMS

FAX	Facsimile
FCC	Federal Communications Commission
FF	Fast-Forward
FMEA	Failure Modes & Effects Analysis
GUI	Graphical User Interface
GSM	Global System for Mobile Communications
HD	High-Definition
HDMI	High-Definition Multimedia Interface
HIC	Head Impact Criteria
HMI	Human Machine Interface
IAPA	International Airline Pilots Association
IFE	In-Flight Entertainment
IFEC	In-Flight Entertainment & Connectivity
INCAM	Integrated Noise Cancelling Audio Module
ISP	Internet Service Provider
ISPS	In-Seat Power Supply or In-Seat Power System
Ka Band	Frequencies between 26.5GHz and 40GHz
Ku Band	Frequencies between 12GHZ and 18GHz
LAN	Local Area Network
LCD	Liquid Crystal Display
LRU	Line Replaceable Unit
MAC	Media Access Control
MAS	Matsushita Avionics Systems
MMS	Multimedia Messaging Service
MTBF	Mean-Time-Between-Failure
MUX	Multiplexed
O.D.	Outside Diameter
OEM	Original Equipment Manufacturer
OLED	Organic Light Emitting Diode
OS	Operating System
PA	Passenger Address System
PBX	Private Branch Exchange
PC	Personal Computer
PCM	Pulse Code Modulation
PCU	Passenger Control Unit
PED	Portable Electronic Device
PES	Passenger Entertainment System
POI	Point-of-Interest or Position-of-Interest

ACRONYMS

PRAM	Pre-Recorder Announcement Machine
PSS	Passenger Service System
PSU	Passenger Service Unit
QoS	Quality of Service
RDU	Receiver Decoder Unit
REW	Rewind
RF	Radio Frequency
RNCU	Radio Network Control Unit
SATCOM	Satellite Communications
SEB	Seat Electronics Box
SD	Standard Definition
SDI	Serial Data Interface
SMS	Short Message Service
STC	Supplemental Type Certificate
TSO	Technical Standard Order
TTL	Taxi, Takeoff and Landing
TV	Television
UHDTV	Ultra-High Definition Television
US	United States
USB	Universal Serial Bus
USD	United States Dollars
VCC	Video Control Center
VCR	Video Cassette Recorder
VHS	Video Home System
VIP	Very Important Person
VVIP	Very Very Important Person
WAEA	World Airline Entertainment Association

Briski Consulting, LLC

We are successful at partnering with core business operations to significantly expand market share, generate sustainable revenue and grow profits. We provide strong domestic and international management experience with a track record of building high-performance, multi-cultural teams.

We have the proven ability to manage product development/production processes from concept through to finished product, with extreme attention to budgets, schedules, quality and specifications. We provide extensive international experience and well-honed presentation and negotiation skills within a multi-national environment.

Key Areas of Expertise Include:

- Corporate Vision & Operational Planning
- Bidding/Negotiating Large Contracts
- Global Market Penetration & Pricing Strategies
- Forecasting & Budgeting
- Strategic & Tactical Planning
- Team Building & Leadership
- ROI, Value Proposition & Profit Optimization
- Conference & Trade Show Presentations
- New Product Architecture & Specifications
- New Product Introduction & Launches
- Analyzing Emerging Technology Trends
- Program/Project Management
- Product Refinement & Life-Cycle Management
- Product and System Certification Support
- Writing and Implementing Sales & Marketing Plans
- Speaking Engagements & Media Relations

Contact us for help expanding your market share and for getting your business and/or product development back on track.

www.BriskiConsulting.com
Info@BriskiConsulting.com

Made in the USA
Las Vegas, NV
10 November 2026

57843101R00142